Wheelchair housing design guide

Second edition

Stephen Thorpe
and
Habinteg Housing Association

HABINTEG
HOUSING ASSOCIATION LTD

Authors' details

Stephen Thorpe
Threshold Architects
72 Friars Street
Sudbury
Suffolk
CO10 2AJ
Tel: 01787 881661

Habinteg Housing Association Ltd
Holyer House
20–21 Red Lion Court
London
EC4A 3EB
Tel: 020 7822 8700
Website: *http://www.habinteg.org.uk*

BRE Press and Habinteg Housing Association make every effort to ensure the accuracy and quality of information and guidance when it is published. However, we can take no responsibility for the subsequent use of this information, nor for any errors or omissions that it may contain.

BRE Press supplies a wide range of building and construction related information products from BRE and other respected organisations.

Details are available from:

www.brepress.com
or
IHS Rapidoc (BRE Press)
Willoughby Road
Bracknell RG12 8DW
Tel: 01344 404407
Fax: 01344 714440
email: brepress@ihsrapidoc.com

Requests to copy any part of this publication should be made to:

BRE Press
Garston, Watford WD25 9XX
Tel: 01923 664761
email: brepress@emap.com

Photos courtesy of: Jacky Chapman/Photofusion (cover top left, page 1 left); Kenneth C Zirkel/iStockphoto (cover top right, page 1 right); Easy Living Home info@easylivinghome.co.uk (cover bottom, page 17); Brenda Prince/Photofusion (page 1 centre).

Index compiled by Linda Sutherland

EP 70
© Habinteg Housing Association 2006
First published 2006
ISBN 1 86081 897 8

Contents

Foreword

Wheelchair housing can seem such a strange term – housing for wheelchairs?

Well, I'm a full-time wheelchair user and yes I and my wheelchair have a home! Of course, all of us have our own housing stories: first homes, renting, buying, sharing with others and moving to other areas; but for wheelchair users options are often very limited and the stories of the difficulties of ever finding anything suitable distressingly familiar.

My own housing story starts in this way, trying to find any home in London that could meet the needs of a young man moving to the Capital to start his career. In the 1970s this proved extremely difficult but I was lucky, I found a specialist housing association, Habinteg, the commissioner of this guide.

By now I would have hoped things have improved and that a young disabled person today would have greater opportunity to find a suitable home. Unfortunately, though some progress has been made, as Chairman of the Disability Rights Commission I still routinely hear stories very similar to my own experience of thirty years ago.

This guide contains the wisdom of many years of developing housing for wheelchair users. It forms part of the ambition to develop truly accessible environments that routinely include and meet the needs of individuals rather than creating barriers that exclude.

Much of the work in developing this guide comes from the feedback of wheelchair users living in such homes, mainly built by housing associations and local authorities.

More homes like these are needed not just to house those who are currently living in unsuitable housing but also to ensure that disabled people have the same sort of choice as non-disabled people. It means ensuring that the principles within the guide are adopted by all developers of new housing, in both the public and the private sector.

Bert Massie

Chairman, Disability Rights Commission

January 2006

Preface

I welcome the publication of the second edition of the *Wheelchair housing design guide*. Housing associations have a long history of ensuring that housing opportunities are available for a wide range of groups. The Housing Corporation recognises this and ensures that part of our programme of investment is geared to meeting the needs of wheelchair users. This guide helps to ensure that the investment we provide delivers quality homes.

Since the publication of the first edition in 1997 we have seen the creation of the Disability Rights Commission, the revision to Part M of the Building Regulations and the adoption of 'lifetime homes' within many strategic planning strategies. This technical guide forms part of this movement to create housing opportunities for all with a clear focus on promoting disability equality. The Housing Corporation is fully committed to this task and will endeavour to ensure that our public investment supports the creation of homes and communities that are accessible to all.

Jon Rouse

Chief Executive, Housing Corporation

January 2006

HOUSING CORPORATION

Background

The first edition of this guide was commissioned by NATWHAG, the National Wheelchair Housing Association Group, and Home Housing Trust, with a steering group on which both were represented together with the Housing Corporation.

Research was undertaken amongst a cross-section of wheelchair users, and their original profiles are reproduced in this second edition.

Research was also undertaken amongst housing association practitioners, and some testing at a late stage in the guide's production was undertaken by architects and other professionals working in this field.

This second edition was revised by the original author and Chris Goodman, Design and Development Manager at Habinteg Housing Association. Feedback from users of the first edition was obtained by the publishers, and key issues raised in the preparation of the early draft of this edition were aired and discussed at a symposium with participants representing the Housing Corporation, OPDM (Office of the Deputy Prime Minister), BRE, Joseph Rowntree Foundation, Centre for Accessible Environments, architects, access officers, occupational therapists and housing associations.

Acknowledgements

We are grateful to the following people who gave valuable advice in the preparation of this second edition or provided feedback on the first edition.

Symposium participants/consultees
Chairman: Mike Donnelly, Habinteg
David Bonnett, David Bonnett Architects
Caitriona Carroll, Joseph Rowntree Foundation
Lawrence Chee, Housing Corporation
Nick Clarke, BRE Press
Clive Clowes, Housing Corporation
Tim Fallon, John Grooms
Julie Fleck, London Development Agency & Greater London Authority
Paul Gamble, Habinteg Housing Association
Bronwen Manning, David Bonnett Architects
Kenneth Matheson, London Borough of Hammersmith and Fulham
Peter Nicholson, Edwin Trotter Architects
Clair Parfrey, Centre for Accessible Environments
David Petherick, Office of the Deputy Prime Minister
Kate Sheehan, College of Occupational Therapists
Edwin Trotter, Edwin Trotter Architects
David Williams, Ability Housing Association
Mike Wilton, John Grooms
Mike Wright, BRE

User feedback
Jeremy Beal, Marches Housing Association
Peter Bowen, Thames Valley Housing Association
Norman Cash, Islington & Shoreditch Housing Association
Jane Lowe, Staffordshire Housing Association
Kate Mallett, Buschow Henley
Sean McKeon, Faithful & Gould
Ian Powell, Tremain Powell Partnership
Paul Timmins, Stafford County Council

Introduction

Since the first edition of this guide was published in 1997 there have been significant developments within the housing field, or affecting it, which have made necessary a revised edition that reflects more accurately the current context.

The purpose of the guide remains unaltered – to explain how to design and detail a home that is fully manageable by wheelchair users and maximises their independence.

It remains activity based, informed by the original research and by the experience of the individual wheelchair users, whose profiles are retained. It takes account also of feedback from practitioners who have used the original guide.

The most significant development has been the increasing acceptance of the 'lifetime home' concept from which housing designed specifically for wheelchair users as promoted by this guide should be clearly distinguished.

The principle underlying the lifetime home and the standards which define it is that such a home should be basically accessible but should incorporate built-in or designed provision for a range of adaptations that would respond to the needs of any member of the household who has or develops a disability or impairment which could include the need to use a wheelchair within the home. These potential adaptations range from enabling access between floors by means of a stairlift or through-floor lift, to hoist installations and provision of a floor-level shower. The lifetime home standards extend the guidance included in those sections of *The Building Regulations 2000 Approved Document M* (2004 edition) which deal with housing.

These two approaches to housing are increasingly being brought together in local plans which establish the lifetime home as the basic requirement for general needs housing with the inclusion of a percentage of fully wheelchair accessible homes.

This development widens both the provision of accessible housing – by private developers as well as social housing organisations – and the scope of such provision. Whereas both wheelchair user and lifetime homes have traditionally been single- or two-storey houses on appropriate sites, they are now being incorporated in multi-storey developments as apartments or duplexes where there is a different range of considerations on accessible design, as defined both by lifetime homes standards and by this guide, which are not found in low-rise housing. Such considerations include fire-resisting doors within the dwelling, the construction of intermediate floors and the need to integrate wheelchair user housing without either compromising the essential standards set out in this guide or exceeding economic space standards for the other dwellings. Furthermore, the demand to maximise inner city sites is leading to unorthodox solutions, including conversion of existing non-residential buildings.

Format

This guide explains how to design and detail a home for wheelchair users. The emphasis is on usability. Because it is activity based, it makes designers think carefully and design confidently and inventively; it does not, therefore, include plans or preferred solutions. Each activity is introduced with an overriding *Principle* of usability; then the implications of this principle are discussed under *Design considerations*. Specific guidance is set out in the form of *Requirements* and *Recommendations* which are summarised in checklists in the Appendices. *Requirements* state essential design, and *Recommendations* set out the best practice in the design of wheelchair user housing. No-one who uses a wheelchair should, as a result of the way their home is designed, be restricted in their independent use of it. Nor should they experience indignity, unreasonable discomfort or inconvenience in carrying out everyday activities.

Scope

The primary aim of this guide is to ensure the good design and detail of new-build homes, or homes provided through rehabilitation or conversion, that are intended to be fully accessible to wheelchair users. It will also be valuable in the design of:

- general housing whether or not it is designed to lifetime homes standards

- the one-off house designed around an individual wheelchair user who may have significantly more demanding needs and the necessary financial resources. In this case it will provide a base brief

- specific adaptations to suit a wheelchair user, although it is not intended in other respects as a guide to house adaptations

Basis

The human considerations on which this guide is based are as follows.

- Independence in managing domestic activities, particularly personal ones, is highly prized. Assistance where it is provided is never continuous, and design should enable the highest degree of self-management.

- There is no typical wheelchair user, and those who need to use a wheelchair in their own homes represent a cross-section of the population. They include young and grown-up children in families, parents (some of them single), people living in partnerships and people living alone.

- Individuals are tending to use more than one wheelchair, manual or electric. Some people use larger outdoor vehicles such as scooters. Wheelchairs vary in size and type, as do their users' abilities to control and manoeuvre them independently.

- The reasons why someone needs to use a wheelchair may have a bearing on their other physical and sensory capabilities. They may be able to reach only a short distance up, down or across. Even then, they may need the support of a fitting or their other arm. They may not be able to reach backwards to a door handle or light switch. They may find it difficult to manage controls and fittings that require precise finger location, the use of both hands or complex manipulation. They may need good visual or audible information. Needs may change through ageing or other factors.

- A person's perception or expectations of their home, or their desire to enhance it, is not diminished because they need to use a wheelchair within it. Good design should seek to avoid such potentially negative aspects as:
  - damage to walls and doors,
  - floor surfaces that are difficult to clean,
  - exposed pipework under sinks, and
  - bathrooms or kitchens of an institutional character.

Adaptations

Housing intended to suit a range of wheelchair users may need to be adapted to:

- allow for a person's changing needs or capabilities within their present home

- allow reasonable compromise to be achieved between the wheelchair user and other household members

- suit successive occupants of a house, some of whom may have quite specific and different needs

- incorporate improved standards of communication, security and safety

Providing specialist components that enable a high degree of adaptability to suit the widest range of potential requirements and significantly increase initial costs, may not be appropriate or justified.

On the other hand, some provision to allow fine tuning or adaptation is desirable. Past experience, and research into future needs and developments, can indicate what is likely to be commonly required. If provision is carefully incorporated at the outset it may involve little extra cost. It may also avoid or limit expensive or disruptive subsequent work.

The *Design considerations* discuss such provision and the *Requirements* and *Recommendations* incorporate it where it is considered appropriate or practicable. Particular examples are additional door fittings, support rails to walls, or additional controls.

Specific adaptations, whatever the provision made to facilitate them, are not the concern of this guide.

Legislation and Standards

Disability Discrimination Act 1995

The provision of housing as a service comes within the scope of the Act but the duties of service providers which can have a bearing on the design of the buildings open to or used by the public or used as places of employment, do not currently extend to the design of housing.

British Standard BS 8300:2001

This Standard (Design of buildings and their approaches to meet the needs of disabled people) published in 2001 considerably expands the information in BS 5810:1979 (Code of practice for access for the disabled to buildings) and BS 5619:1978 (Code of practice for design of housing for the convenience of disabled people), which it replaces. The design of housing is not specifically addressed, but there is detailed guidance on a range of elements, such as parking, design of ramps and stairs, common areas, lifts, doors and controls, which are relevant to this guide and are incorporated where appropriate. The detail design of lavatories and bathrooms relates to 'buildings that require bathing facilities' which would include communal facilities in supported housing. There is a suggestion in the Standard that it 'may be applied to dwellings for individuals' and this has been followed where appropriate.

Approved Document to Part M of the Building Regulations

The 1999 edition included a separate section on the design of new housing. The 2004 edition takes account of BS 8300 for the design of buildings used by the public, including alterations and extensions and common areas of housing developments, but the sections specifically devoted to the design of new housing are not wider in scope than the 1999 edition. They provide guidance on achieving a basic accessibility – approach and entry into the dwelling, circulation within the entrance or principal storey, provision of a nominally accessible WC at this level, the design of the staircases, and heights of switches and sockets. There is an ongoing discussion regarding a further edition which will incorporate more of the present lifetime homes standards but will not extend to specifically wheelchair housing.

Design for wheelchair users

Basic design considerations

The basic design considerations set out below should enable designers and planners to make the appropriate early decisions which ensure that the detail design can follow the guidance set out in the body of this guide.

These design considerations, and those contained within the main body of the guide, on an initial read may seem to make greatly increased space demands. However, the design process, following all the issues, can and should lead to solutions that optimise the extra space required and reduce the need for a large footprint.

As further guidance, annotated plans which demonstrate typical solutions for a range of dwelling types and indicate how the pattern of circulation and relationship of rooms and spaces affects the overall footprint, are posted on the Habinteg website (*http://www.habinteg.org.uk/whdg*).

Inclusion

As an overriding principle, aim at inclusive solutions, avoid differentiating by location, form and detailing, and in multi-storey developments provide a range of choice – size, aspect, floor level.

Siting

The initial selection of the site should have regard to how wheelchair users will gain access to facilities beyond it and the transport options available to them.

Ensure good inclusive and convenient access for wheelchair users throughout the development, linking entrance to site, all entrances to dwellings and site facilities, and resolving differences in level so that detailed design guidance – gradients, widths, crossings, cross falls, surfaces – can subsequently be followed.

Resolve access by car, parking strategy, approach to entrances from car parking and pedestrian network, having regard to distance between parking and dwelling, need for nominally level parking, and for covered transfer between car and wheelchair where required.

Resolve potential need for scooter storage, charging and covered transfer adjacent to entrance.

Ensure that any covered transfer space for car or scooter blends with the overall design of the scheme.

Ensure that covered space at entrance is a functional porch or canopy rather than a decorative one.

Internal planning

Ensure accessible circulation (widths and turning spaces) throughout communal areas on ground floor and other floors served by lifts.

Ensure that domestic lifts in two- and three-storey houses link circulation areas at each level.

Ensure access to a minimum of two lifts in multi-storey developments with preferably one of at least 12 persons capacity.

Ensure efficient circulation that will enable a wheelchair user to turn 180 degrees at the front door, and approach/turn through all doorways, around corners and approach built-in cupboards/storage.

Ensure dedicated space within the dwelling which will enable a wheelchair user to manoeuvre and transfer to a second wheelchair, leaving the other wheelchair on charge without it interfering with circulation space.

Ensure as far as practicable that all doors open beyond 90° to achieve effective clear width of opening without excessive door width.

Determine relationship or differentiation of living and dining areas, kitchen and circulation, taking account of household type, dwelling size and likely living patterns, ensuring ease of wheelchair movement and providing some flexibility in use, such as scope to screen off or enclose the kitchen.

Ensure that living areas can be satisfactorily furnished, and accommodate wheelchair movement, including access to windows and external doors.

Ensure that kitchens can be used conveniently by wheelchair users, **take account** of need for knee space under sink, hob and some worktops, access to oven, other appliances and storage, **make** practical and functional provision for height adjustment, **minimise** need for excessive movement between kitchen functions, ie L- or U-shaped layouts rather than linear ones and with convenient access to dining area.

Provide at least one double and one single bedroom with provision for direct connection between main bedroom and bathroom.

Ensure that a wheelchair user can enter and leave all bedrooms easily, access both sides of a double bed or one side of a single bed, turn and manoeuvre to approach all furniture and operate window controls. *Refer to overall layout in Section 12.*

Ensure that bathrooms can accommodate a level-access shower area at least 1000 × 1000 mm, without an approach ramp, overlapping a 1700 mm bath area to allow for either to be installed; allow space for side, oblique and front transfer to the WC, a 1500 × 1500 mm manoeuvring space clear of all fittings, 1100 mm between inward-opening door swing and WC, and provision for direct connection from adjoining main bedroom. *Refer to overall layout in Section 11.*

Provide in larger dwellings a separate WC, possibly with provision for a shower in the side transfer space, with WC handed opposite to that in the main bathroom.

Provide ample appropriate storage throughout with a good degree of this accessible.

Ensure unhindered access to operate windows and approach secondary doors, and fit remote control devices where, for example, windows are above kitchen units or bathroom fittings.

Management

Clarify and resolve issues such as evacuation from multi-storey developments, delivery and refuse arrangements for residents and visitors.

Adaptability

Plan for life of building not just initial occupants, and aim for a degree of built-in adaptability.

Outline of design guidance

The technical sections which follow deal with the range of activities related to the domestic environment. They comprise 15 activities which are grouped and ordered in a sequence that is user oriented. Designers may wish to consider them in a different sequence but will find all the information they need.

External environment and entrances

1	Moving around outside
2	Using outdoor spaces
3	Approaching the home
4	Negotiating the entrance door
5	Entering and leaving; dealing with callers
6	Negotiating the secondary door

Internal environment

7	Moving around inside; storing things
8	Moving between levels within the dwelling
9	Using living spaces
10	Using the kitchen
11	Using the bathroom
12	Using bedrooms

Components and details

13	Operating internal doors
14	Operating windows
15	Controlling services

Each of the 15 activities comprises *Principle, Design considerations, Requirements* and *Recommendations.*

- **Principle** sets out the realistic degree of usability to be aimed at

- **Design considerations** discusses the implications of applying the *Principle* to layout, design, and detailing, with explanatory illustrations

- **Requirements** set out what are considered as the minimum essential standards for all housing for wheelchair users

- **Recommendations** set out what are considered as desirable features which should be met by all providers of housing for wheelchair users wherever practicable

In both *Requirements* and *Recommendations* the illustrations are an integral part of the guidance.

The technical sections conclude with the following Appendices:

- Summary of *Requirements*

- Checklist of best practice

(These are clearly cross referenced to specific *Requirements* or *Recommendations*. They do not stand alone and are only to be referred to when the *Design considerations* have been studied, the *Requirements* met and as many of the *Recommendations* incorporated as is practicable.)

Note for affordable housing designers and providers

The Housing Corporation, when assessing accessibility standards for wheelchair housing, will have regard to the *Requirements* as described in the technical sections of this guide.

Note for all designers and developers

At times the numerous dimensional requirements and recommendations give the impression that footprints of dwellings or particular rooms must carry an onerous cost due to their required size. However, with careful design, consideration and thought on how activity spaces can be multi-functional, in many cases the requirements and recommendations can be achieved within reasonable overall dimensions. Increased space is always welcome and desirable – it increases choice and eases manoeuvre for wheelchair users, but this must be tempered by what is realistic and affordable within cost restraints. An example of this is bathroom layout (see Section 11). A well designed bathroom is key to enabling independence and dignity within the home, and the dimensional requirements and recommendations to enable a suitable layout are numerous. However, Figure 11.1.5/11.1.7 shows how these can be incorporated efficiently within an enhanced, but reasonable, overall area.

Using a wheelchair

It is important that designers have a clear understanding of how wheelchair users negotiate the external and internal environment.

The wide range of manually and electrically operated wheelchairs and of the techniques adopted by those who use them means that it is difficult to be specific about the space needed within the home for turning and manoeuvring.

The familiar 1500 mm diameter circle may be used as a rule of thumb at the planning stage, provided it is realised that because the axis of the manual wheelchair is off centre, a full turning circle would be rather larger. The smaller circle is in practice only of practical value if there is usable space outside it to assist manoeuvre, such space typically being provided between or beneath fittings or by a carefully positioned opening.

It may be helpful to note that BS 8300, for which research into the spatial requirements of wheelchairs in use was undertaken, adopts in some of its diagrams the 1500 × 1500 mm turning square. This reflects the fact that manoeuvring a wheelchair often includes several moves rather than the smooth or complete turn implied by the circle.

It is also important for designers to appreciate how a wheelchair user approaches, operates and passes through an internal doorway.

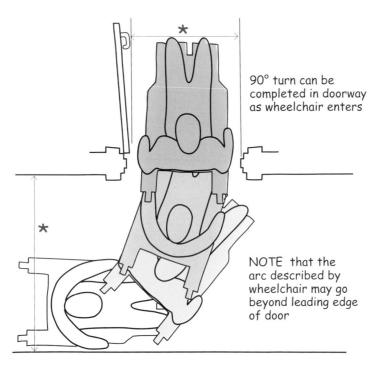

90° turn can be completed in doorway as wheelchair enters

NOTE that the arc described by wheelchair may go beyond leading edge of door

* For importance of passage width in relation to effective clear width of doors refer to Section 7

It is obviously not possible for designers to place themselves in the position of the regular and individual wheelchair user. Nevertheless, some basic hands-on experience of using a manual or electric wheelchair to supplement close observation would be invaluable. Local access groups may run awareness courses or make chairs available.

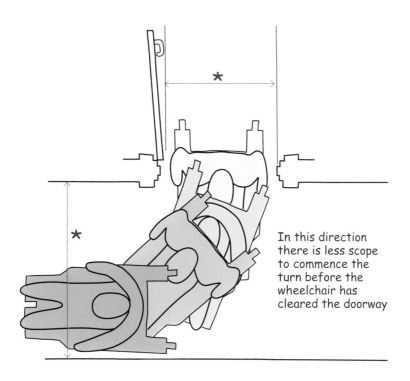

In this direction there is less scope to commence the turn before the wheelchair has cleared the doorway

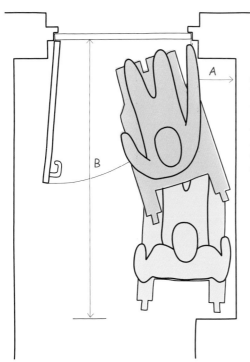

Space at A beside the door enables the wheelchair user to approach and release the door, then to reverse while holding the door, until the wheelchair is clear of the door swing, then to move forwards and pass through the doorway

An increase at A beyond the required 300 min will reduce the space needed to clear the door swing and thus reduce dimension B

Dimensions and conventions in this guide

In both text and illustrations dimensions are always in millimetres and are in accordance with the conventions set out below.

Horizontal

Dimensions are between or from finished wall surfaces.

Vertical

Dimensions are from finished floor or paving surfaces.

Fittings, controls

Dimensions are to centre line of fitting, or control, or to centre line of a particular component.

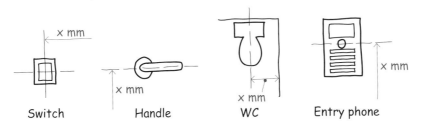

| Switch | Handle | WC | Entry phone |

Fittings, controls

Doors, basic criteria

Positioning

Positioning dimensions are to nearest edge of door from finished wall surface.

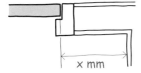

Widths

BS 8300 redefines clear opening width as effective clear width to take account of projections from the face of the door such as door handles and weatherboards. Observation of this principle in domestic planning may result in excessive actual door widths – ensuring that a door can open beyond 90° may be economical of space and cost and make for easier circulation, as discussed in Section 13 *Operating internal doors*.

Effective clear width is between face of door or projecting fitting in open position and nearest point on opposite frame or second door.

The relationships between doorset, door and effective clear width for the different door types referred to in this guide are illustrated below.

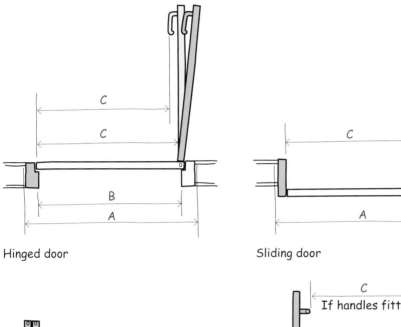

Hinged door

Sliding door

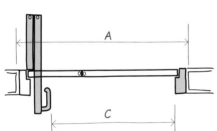

Reduced-swing door

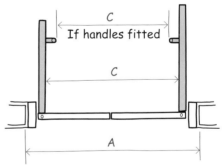

Paired hinged or swing doors

In the illustrations:

A = doorset width, ie door + frame or lining within structural opening

B = actual door width

C = effective clear width available for passage through doorway

These dimensions indicate only the relationship of A, B and C – refer to Sections 4, 6, 7 and 13 for detailed guidance on door dimensions.

Layout of rooms, basic criteria

Rooms should be laid out to provide adequate space for a wheelchair user to move between furniture and fittings. The clear zones given below should be incorporated into notional furniture layouts in order to determine workable room sizes and proportions.

Refer also to:

- Section 7 *Moving around inside; storing things*
- Section 9 *Using living spaces*
- Section 12 *Using bedrooms*
- The National Housing Federation's *Standards and quality in development: a good practice guide* (see Bibliography).

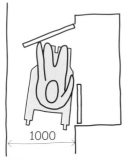

1000

Sideways approach to wardrobes, storage

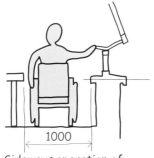

1000

Sideways operation of windows

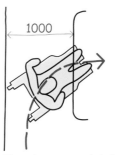

1000

Turn to approach tables with knee space under

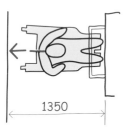

1350

Space to approach, reverse, pull out drawers

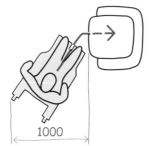

1000

Space to transfer to chair

Clear zones

800 mm   clear passage between furniture

1000 mm   clear width to ensure sufficient space for:

- sideways approach to wardrobes, storage
- sideways operation of windows, for example
- side approach/transfer to one side of beds (see Section 12 for space required to move around foot of bed to approach the other side)
- turning and approach to dressing tables and dining tables with clear space under for footrests
- transfer to armchair or settee

1350 mm   space to approach furniture and reverse to pull out drawers

Hoists, design criteria

Hoist

A ceiling-mounted tracked hoist if subsequently installed will be set out to suit a specific user, but the following aspects are relevant to making provision within the bathroom and bedroom layouts considered in this guide.

Layout

A track may be straight, possibly at an angle to the room, or it may be curved, or incorporate junctions. It may connect rooms, adjacent rooms or spaces or serve particular operations such as transfer from wheelchair to bath, bed or seating, or transfer between wheelchairs.

Any built-in strengthening should therefore allow for the full range of possible installations.

Track/door head

A connecting door when the hoist track is installed must extend to the latter and allow the traversing mechanism to pass through the opening. Where the mechanism overlaps the track, doors must be paired and notched. In other cases doors may be single or paired. Sliding doors will always need to be paired with separate tracks.

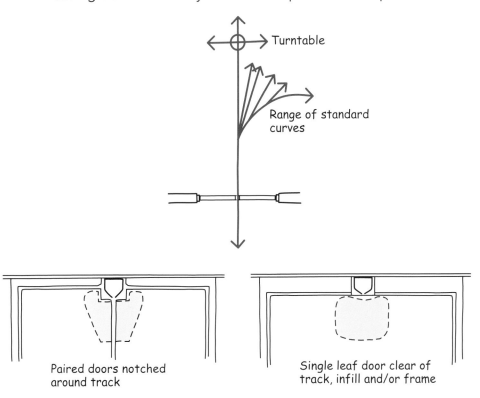

Paired doors notched around track

Single leaf door clear of track, infill and/or frame

In using the guidance, designers for Registered Landlords and the Landlords themselves should refer to the notes for designers on page 9.

Technical sections

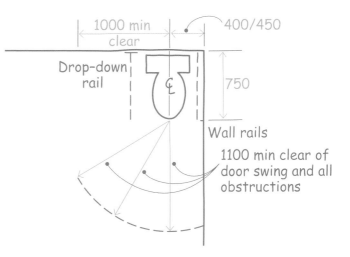

1 Moving around outside

Principle

Ensure a high degree of accessibility within the development.

1.1 Design considerations

1.1.1 Identification of suitable sites should take account of the considerations which follow.

1.1.2 Movement off site is easier for wheelchair users who are drivers or regular car passengers or where there is some form of accessible transport. Those who rely on a manual or electric wheelchair or scooter will define facilities 'within reach' in more restricted terms. In their case the quality of detailing will determine realistic travel distances by such means. The policy and record of the local authority in maintaining paved surfaces and lighting or in extending accessibility to crossings by dropped kerbs or raised roadways will be a significant factor. It is therefore important to minimise the adverse impact on accessibility of local authority adoption standards.

Figure 1.1.2

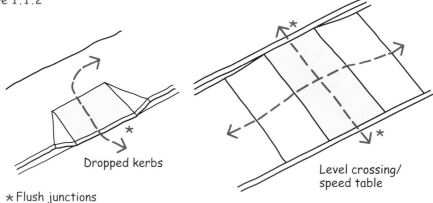

Dropped kerbs

Level crossing/
speed table

★Flush junctions

1.1.3 Basic facilities are the same for everyone: shops, post office, post box, cash point, schools, doctor, chemist, pub. Easy access to local shops by wheelchair is rarely feasible and in many places, for lower prices, convenience and accessibility, the tendency is towards bulk supermarket shopping by means of car or taxi.

1.1.4 For movement within the site, eg for resolving levels and surface drainage, the independent use of a manual wheelchair should be the criterion.

1.1.5   Routes within a development should be fully accessible, safe and secure. This implies careful layout and detail to resolve differences in level, crossings, incorporate overlooking, lighting and signing and to eliminate hazards.

1.1.6   Slopes and gradients are necessary at changes in level but they can inhibit ease of movement and should, where practicable, be eliminated or kept short and to shallow gradients.

1.1.7   Cross falls along routes, ie at right angles to wheelchair passage, can present difficulties. A manual wheelchair has a tendency to drift off course and a powered one to skid. Such falls should be eliminated or at least kept to a minimum gradient. Similarly, falls in two directions should be eliminated where possible or kept to minimum gradients.

Figure 1.1.7

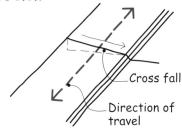

Cross fall

Direction of travel

1.1.8   Some wheelchair users are sensitive to rough surfaces. Selection and detailing of paving should ensure that surfaces are laid and remain smooth (unless designated hazard paving).

1.1.9   Some wheelchair users within their homes may feel particularly vulnerable where their windows open onto routes or undefined spaces. Where front gardens are not provided, defensible space should be established by low walls, fences, robust planting. See Section 2 *Using outdoor spaces* and *Secured by design* (page 115).

Figure 1.1.9

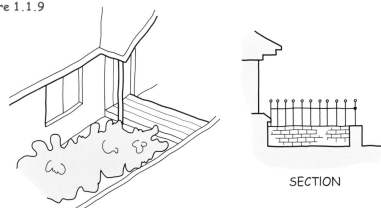

SECTION

1.1.10 Single-storey wheelchair homes at ground level are usually the most easily designed and can be cost-effective. Other possibilities such as two- or three-storey dwellings or apartments or flats should be considered or may be an outcome of local planning policies and negotiations.

1.1.11 The distinctiveness of wheelchair housing should be minimised. Some means of achieving this are: sensitive treatment of garaging and parking, and common detailing of features such as level thresholds, canopies and entrance doors, brickwork and roof tiling.

1.1.12 All dwellings on a development should be accessible to a visitable standard; a wheelchair user intending to visit friends and neighbours should be able to approach, call at and enter the home and use at least its main living area and entrance-level WC. This should be achieved in new housing built to guidance in the 2004 edition of Building Regulations Approved Document M, but the fully accessible WC will generally only be found in flats or larger houses (with three or more bedrooms) designed to 'lifetime homes' standards.

1.2 Requirements

1.2.1 Footpaths

Ensure that footpaths are smooth but slip resistant, of 1200 mm minimum width and with adequate space to negotiate obstacles, turn and pass.

Figure 1.2.1

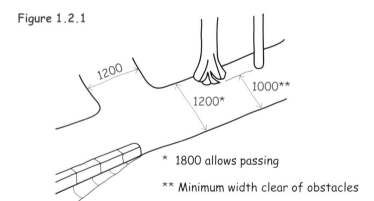

1200

1200*

1000**

* 1800 allows passing

** Minimum width clear of obstacles

1.2.2 Footpath edges

Provide protective edgings, kerbs or rails where footpaths are significantly higher than adjacent ground levels or where adjacent ground significantly falls away.

1.2.3 Gradients

Ensure that gradients to footpaths or routes within a development do not exceed the distance or ratio shown below. Ensure that maximum length of slope is limited to suit the gradient with top, bottom and, where required, intermediate landings, all as illustrated.

Figure 1.2.3

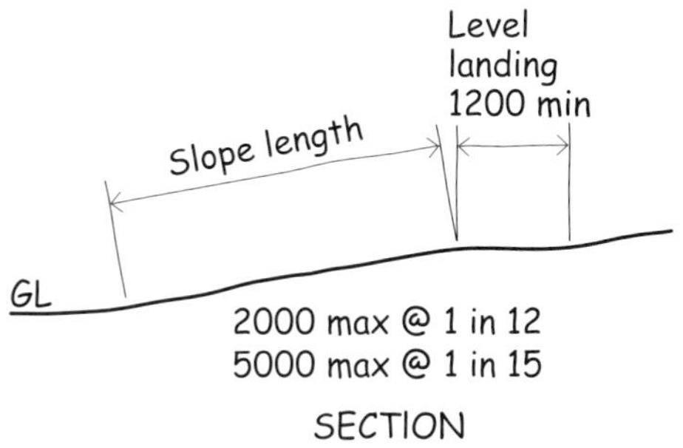

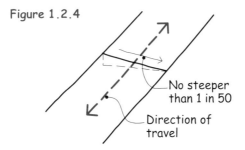

SECTION

1.2.4 Cross falls

Ensure that these do not exceed 1 in 50 on paving, whether dedicated to pedestrian use or shared with vehicles.

Figure 1.2.4

1.2.5 Crossings

Ensure that these have flush junctions or shallow gradients and avoid gratings or channels which could trap wheels or footrests.

Figure 1.2.5

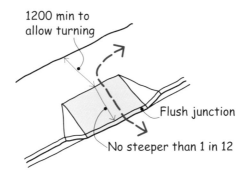

1.3   Recommendations

1.3.1   Gradients

Ensure that gradients to footpaths or routes do not exceed
1 in 20.

1.3.2   Cross falls

Ensure that these are avoided wherever possible.

1.3.3   Routes

Ensure that routes are safe, avoiding obscure or out-of-view
sections, with a consistent and good level of illumination.

1.3.4   Frontage

Ensure that there is a defined space in front of windows facing
onto routes.

1.3.5   Appearance

Ensure that wheelchair user homes by their siting and
appearance are a natural part of development and not
excessively distinctive.

1.3.6   Entrances

Ensure that these are identifiable from routes by signing and
lighting.

1.3.7   Signing

Clarify layout, routes and dwelling numbering system for the
benefit of visitors, following good practice set out in, for
example, the *Sign design guide* by Parker and Fraser (see
Bibliography).

Profile of a wheelchair user 1

This young woman lives alone, with good care support, in a purpose-designed,
housing association, two-bedroom, single-storey house. She has an electric and a
manual wheelchair.

She uses the living room to transfer, with assistance, and to re-charge her electric
wheelchair. She has recently become a car driver. The house has an attached
carport with covered access to the entrance which has an entry phone.

In the ample living space, the woman can reach the window controls if she
approaches them sideways. She doesn't use the kitchen. The worktop, set at
900 mm, is used by the carers. In the bathroom, an automatic wash/dry toilet has
been installed. There is a shower in the designated space. The woman can reach
switches (set at 1080 mm) and sockets (at 890 mm) but not with ease.

2 Using outdoor spaces

Principle

Ensure that spaces within the curtilage of the home are accessible, usable and, where appropriate, secure.

2.1 Design considerations

2.1.1 Front gardens are preferred to other methods of defining the boundary between public and private spaces.

2.1.2 Rear external spaces and gardens should be securely enclosed and overlooked from within the house, particularly where accessed from other than the inside of the house. Where a lock is provided, gates should be capable of being locked and unlocked from each side by a key.

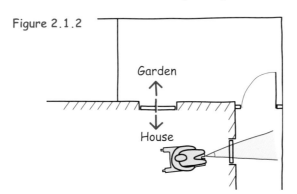

Figure 2.1.2

2.1.3 Where the outdoor space is a balcony or other restricted area it should be possible for a wheelchair user to enter it clear of any external door swings and with internal and external surfaces nominally level. It should also be possible to see over or through the balustrade without danger. Upper-level flats should be provided with balconies, otherwise consider inward-opening doors with balcony front. See also Section 6 *Negotiating the secondary door*.

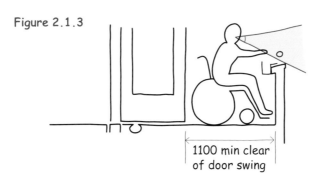

Figure 2.1.3

2.1.4 The garden by its layout should resolve strictly functional requirements such as access to outdoor drying facilities, and routes between home and car, storage and refuse. It should also be capable of being fully enjoyed by a wheelchair user taking account of views from inside the home, orientation and shelter, with scope for accessible paths and for planting.

Figure 2.1.4

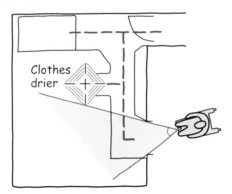

2.1.5 Paving forming paths or sitting-out areas should be properly accessible, ie slip resistant and well detailed to ensure maintained smoothness. Gradients on paths, where unavoidable, should be no steeper than 1 in 20. Sitting areas should be nominally level.

2.1.6 It should be possible independently to deposit refuse and recycled materials in containers for collection. Incorporation within the carport area or under an extended entrance canopy is a possible solution. Management provision may be a determining factor particularly in multi-storey developments.

2.1.7 Outdoor clothes-drying facilities should be usable with a simple means of adjusting the height of the lines.

Profile of a wheelchair user 2

A young woman severely disabled from a car accident. She lives with her parents, sister and a carer in a family-owned extended/adapted bungalow. She uses an electric wheelchair and travels in an adapted vehicle, with tail lift, which backs up to the front door.

Her wheelchair is large and difficult to manoeuvre. She can negotiate a right angle off 1160-mm-wide corridors with the angle splayed, but she needs 1360 mm widening outside her room and a straight approach to the other doors (all 920 mm wide).

In the bathroom there is a hoist track running through from the bedroom. The WC outlet can be adjusted to facilitate transfer from the track.

2.2 Requirements

2.2.1 Gardens

Where private back or side gardens are provided, ensure that gates have 850 mm clear opening, and can be operated from each side by a wheelchair user, with reachable and easily manipulated fittings.

Figure 2.2.1

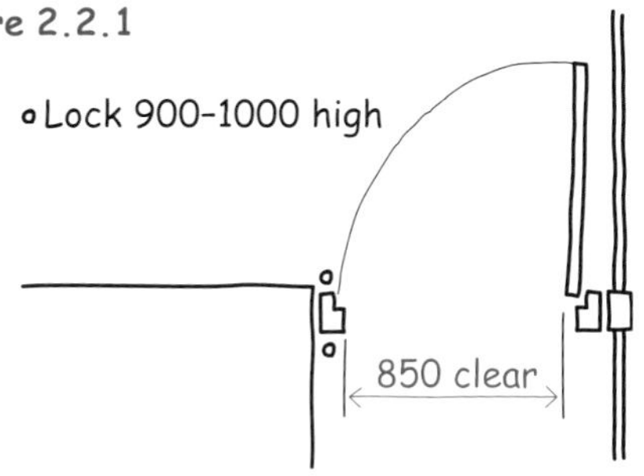

Lock 900–1000 high

850 clear

2.2.2 Balconies

Provide nominally level access to balcony and usable space clear of any door swings.

2.2.3 Garden

Lay out garden to provide:

● accessible paving outside external door

● accessible clothes-drying facilities

● accessible route from external door, external storage and external gate

● scope for accessible planting

2.2.4 Refuse

Make suitable usable refuse provision, ie to deposit refuse in a container for collection, within a short distance of an external door, or ensure appropriate management provision.

2.3 Recommendations

2.3.1 Gardens

Ensure that gardens attached to houses are overlooked from within the home and are fully enclosed, with gates having a clear effective width of 900 mm.

2.3.2 Garden layout

Ensure that gardens may be laid out by occupant(s) with access by paths no steeper than 1 in 20.

Provide some accessible raised planters or beds.

Figure 2.3.2

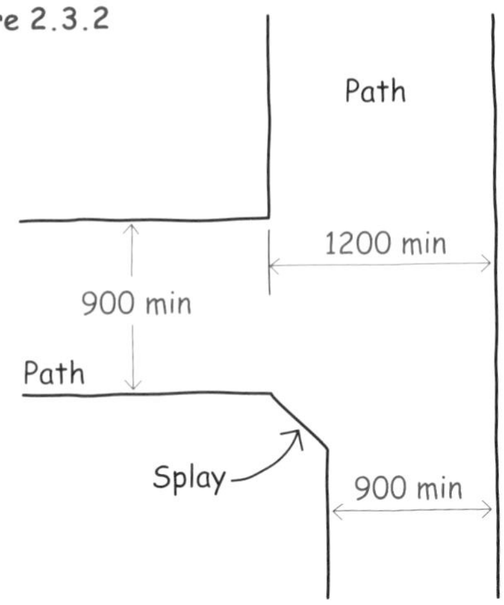

Profile of a wheelchair user 3

A young man lives with his wife and two young children in a purpose-designed, housing association, two-storey house with integral lift. He has one manual wheelchair, and occasionally uses an electric one. He has a regular job and drives a car, so the carport is essential. He drives to the supermarket, as no shops are reachable in his wheelchair.

Although several one- and two-storey houses are designed to be accessible by wheelchair, the remainder of the housing development is inaccessible. The man dislikes this segregation and the vandalism caused by local children. He feels that his car is vulnerable and would like to enclose the carport.

The entrance has an entry phone. He would like external lights fitted with sensors to the front and back entrances. A large matting area helps to clean wheels. He is often visited by friends in wheelchairs.

Circulation is good, with 925-mm-wide doors throughout the house and a minimum passage width of 1170 mm. Circulation spaces are connected by a lift. All bedrooms are accessible. The man values living on two floors, but worries about the reliability of the lift. He can use the lift's integral phone to alert nearby keyholders if he finds himself in difficulty.

The main living space is spacious enough for circulation between furniture. The man is worried about the security of double external doors.

In the kitchen, the worktop is adjusted to a compromise height (835 mm). There is no space for the freezer – he keeps it in the designated wheelchair charging space at the entrance. The floor finish shows wheel scuff marks.

There is a bathroom upstairs and a shower downstairs which he uses without difficulty, although the textured floor is difficult to keep clean. In the bedroom, a hoist track is fitted over the bed.

3 Approaching the home

Principle

Ensure ease of approach to the home by car, wheelchair or intermediate vehicle with good cover at the point of transfer and good protection from the elements at the individual or common entrance.

3.1 Design considerations

3.1.1 The most important need for the wheelchair user is to stay dry, hence weather protection both while transferring outside to a wheelchair and while negotiating the front or common entrance; both operations can be time consuming. This suggests an effective cover over the parking space for car users, a carport for instance and an effective entrance canopy for all. Avoid visually prominent covered routes to entrances, but consider some side protection in exposed settings. A good entrance canopy will also help to ensure the effectiveness of threshold seals in exposed settings.

Figure 3.1.1

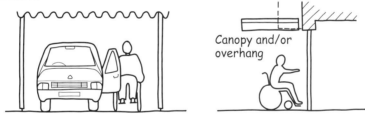

Canopy and/or overhang

3.1.2 It may be possible for the covered area or carport to abut the home and incorporate the entrance canopy. Care should be taken to limit loss of daylight entering the home. Avoid extension of the canopy in front of main windows to living rooms and bedrooms, although small windows looking into carports are very good for car security.

Figure 3.1.2

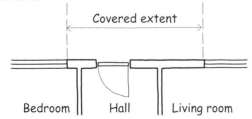

Covered extent

Bedroom   Hall   Living room

3.1.3 Where a garage is not provided, it may be desirable to extend the covered area or carport to provide side protection. Consider siting carports to the sheltered side of the home, or between homes.

Figure 3.1.3

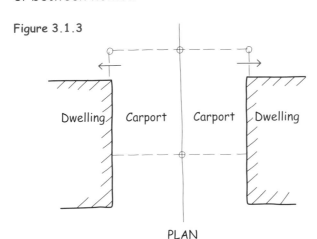

PLAN

3.1.4 A car may need to be driven or reversed in to suit the transfer methods of the wheelchair user, whether driver or passenger, ie transfer from driver or passenger seat or through rear door.

Figure 3.1.4

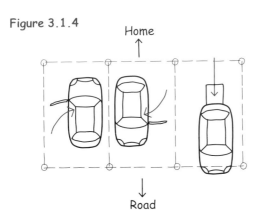

3.1.5 Entry by manual or electric wheelchair into the home may be direct or after transfer from a vehicle. Where the latter is a pavement or class 3 vehicle (scooter) it should be possible to carry out this transfer and store and charge the vehicle within or directly off the covered parking area. Where class 3 vehicle use is probable, external storage with charging facilities should be provided and consideration given to parking and storage.

3.1.6 Grouped car parking serving multi-storey or low-rise high-density developments should be provided on the basis of management arrangements that could provide at least one designated wheelchair space per wheelchair user dwelling. These spaces, whether off street or kerbside, should be of the required size to enable transfer between the car and an adjacent pavement or hardstanding. Wherever practical and feasible, parking should be undercover and travel distances should be minimised. Upstand kerbs should be eliminated as far as possible to maximise scope for transfer and access.

3.1.7 The route to the entrance should be smooth and slip resistant with an effective landing at the entrance. Preferably it should allow for an accompanying adult or child or for a caller using a wheelchair to approach and turn back. In addition some wheelchair users may find a head-on approach difficult and may need to approach the door sideways in order to operate the lock or handle. Easy access to an entry phone call point should be within the landing area.

Figure 3.1.7

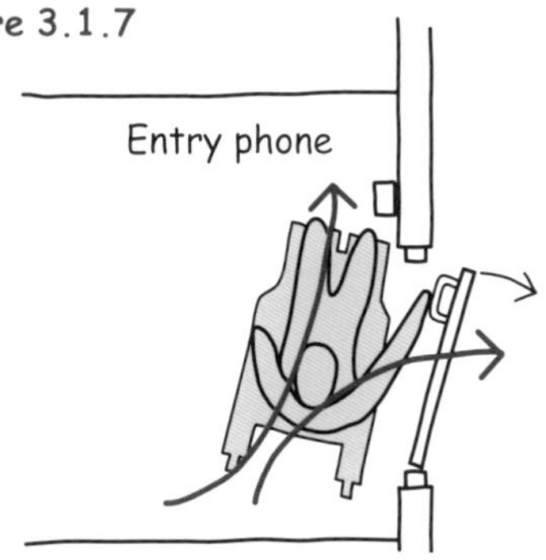

Entry phone

3.1.8 Gradients, ie approaches steeper than 1 in 20, should be avoided if possible. If there is a difference in level between route and adjoining ground level there should be protection as detailed in Section 1 *Moving around outside.*

3.1.9 There should be adequate lighting of a consistent level at the transfer area along the route to and at the entrance, preferably controlled by PIR (passive infrared) detectors.

3.1.10 Garages should provide the same scope for transfer as a carport. Automatic garage door operation would be required.

3.1.11 In common areas of multi-storey developments and others where homes are above the entrance level, follow detailed guidance set out in BS 8300 including circulation areas, doors and lifts. Ensure that two lifts are in regular use, one of which should exceed the minimum eight-person size.

3.2   Requirements

3.2.1   Dwellings with a direct external entrance

Provide a covered parking space for every ground-floor level wheelchair user dwelling.

3.2.2   Covered area

Ensure that minimum clear area and height are as indicated with slip resistant, smooth and nominally level paved surface below.

Figure 3.2.2

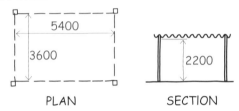

PLAN   SECTION

3.2.3   Dwellings with a communal external entrance

Where there is no direct external ground-floor entrance, ensure parking provision in the form of designated parking space for each wheelchair user dwelling, off street or kerbside, nominally level, that can enable transfer to and from vehicle.

3.2.4   Garage

Garages, owing to their confined nature, are less suitable than carports. Where provided, ensure that minimum area and height are as indicated with nominally level floor surface, and automatic door operation.

Figure 3.2.4

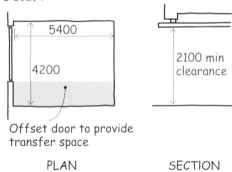

Offset door to provide transfer space

PLAN   SECTION

3.2.5   Route to entrance

Ensure a smooth slip resistant route to entrance. Ensure that ramps, where unavoidable, are not steeper than 1 in 15 and not longer than 5000 mm. Note that it may be possible for the covered area of 3.2.2 to be immediately adjacent to an entrance, providing a direct link and canopy to the entrance.

3.2.6 Entrance landing

Provide level landing 1500 × 1500 mm minimum, but ensure 1200 mm depth clear of any outward door swing. Provide side protection where ground level is below path or landing level.

Figure 3.2.6

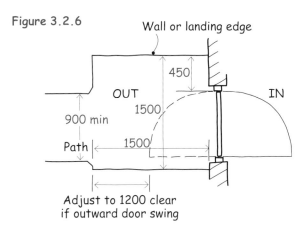

3.2.7 Canopy at entrance

Provide canopy to minimum 1200 × 1500 mm extending beyond door on lock side and at a maximum height of 2300 mm.

Figure 3.2.7

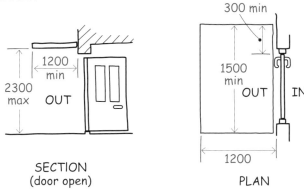

SECTION
(door open)

PLAN

3.2.8 Lighting

Provide adequate lighting of transfer area, route to entrance and entrance itself. Incorporate PIR detectors in external fittings in addition to switching.

3.2.9 Lifts

Where wheelchair dwellings are above the ground floor, lifts should be as detailed in BS 8300. A second lift should be accessible to and from wheelchair user dwellings for use when the first or core lift is undergoing maintenance or is out of service.

3.3 Recommendations

3.3.1 Covered parking

Provide side protection in exposed settings by screening or siting.

3.3.2 Designated parking to flatted developments

Provide cover wherever feasible, blending with the overall scheme design.

3.3.3 Landing

Provide landing 1800 × 1800 mm to suit sideways approach to operate door and access to entry phone.

Figure 3.3.3

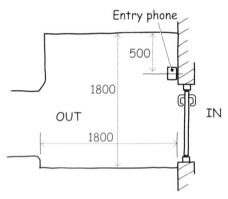

3.3.4 Canopy

Ensure larger canopy than 3.2.7 to provide better cover at entry phone.

Profile of a wheelchair user 4

A young married man sharing an adapted, local authority, ground-floor flat with his wife, who is also a wheelchair user. They are about to move into a purpose-designed house on a small, self-build development comprising five houses. He has been consulting closely with the architect and an occupational therapist.

He has two manual wheelchairs and drives a car. His wife has an outdoor and an indoor electric wheelchair, at present stored and re-charged in the living room. The new house will have an adjoining carport that will incorporate a wheelchair store. He has chosen to omit the partition between passage and living space to maximise the space available for their own and visitors' wheelchairs on entry. Switches are set at 900 mm, sockets at 600 mm.

4 Negotiating the entrance door

Principle

Ensure that any entrance door can be operated and negotiated independently whether the user is entering or leaving. Balance this with optimum security of dwelling whether occupants are out or at home.

Ensure that callers who are wheelchair users can reach and use bell, entry phone and postal plate.

4.1 Design considerations

4.1.1   Main entrance doors include:

- individual external entrances

- common external entrances

- individual internal entrances

4.1.2   An effective clear door width of at least 800 mm will be adequate provided that the passage through in either direction is on line.

4.1.3   Where the door opens towards the direction of approach, the wheelchair user will need to approach the door head-on, release it and reverse while opening the door. This manoeuvre is facilitated by providing a space beside the lock edge of at least 300 mm, preferably up to 550 mm. There should be unobstructed space to reverse the wheelchair clear of the door swing. See *Using a wheelchair* on page 10.

Figure 4.1.3

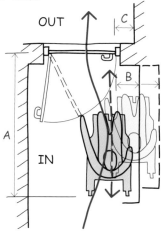

A: space to manoeuvre clear of door swing

B,C: from door edge to wall or landing edge

A reduces as B increases

4.1.4 Where the door opens away from the direction of approach the manoeuvre is simpler, but space beside the lock edge of at least 200 mm will facilitate the approach to release the door.

4.1.5 Thresholds should be selected or detailed to ensure:

- total upstand clear of internal and external finished surfaces of total height not exceeding 15 mm
- tapered profile externally
- weathertightness, taking account of degree of exposure
- adjustability to maintain weathertightness where there are timber doors
- no onerous addition to door opening and closing pressure
- compatibility with installed or added internal floor finishes or coverings

Figure 4.1.5

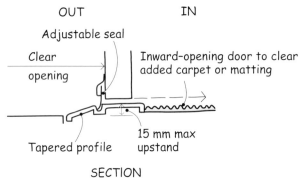

4.1.6 Approach to and operation of door locks can be difficult and it may be necessary for some occupants with limited reach or arm strength to approach sideways rather than head-on. See also Section 3 *Approaching the home*.

4.1.7 The lock may be a component in a remote control door release (entry phone) or door open/close system. Otherwise it should be standard. The optimum standard of security should be achieved.

Profile of a wheelchair user 5

An older woman living with her elderly partner who is also her carer in a one-bedroom, ground floor, adapted, local authority flat. She uses an electric and a manual wheelchair but is unable to use the electric one in the house because space is too tight. She would require an entry phone if she lived alone.

The woman dislikes the 'grottiness' of her shower-only bathroom and the fact that she is unable to see out of her window because it has high sills.

4.1.8 Fittings to open or close doors manually, whether cylinder lock pulls, pulls or lever handles, should all be easy to grip and should contrast with door. Lever handles with mortice latch combined with easily operated multi-locking will provide optimum access and security. Ironmongery should, as far as possible, be consistent throughout the development.

Figure 4.1.8

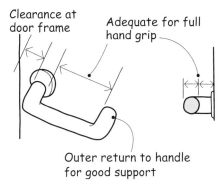

Clearance at door frame

Adequate for full hand grip

Outer return to handle for good support

4.1.9 Common entrance doors, other doors within common spaces and individual entrance doors may all be self closing and fire resisting. Doors fitted with closers can be difficult to operate from a wheelchair. Push-pad door operators may be required to overcome the closing pressure necessary on common entrance doors. For security such operators should be linked to the entry phone installation.

Closers should otherwise be selected and maintained to provide delayed action closing and a maximum opening pressure of 20 newtons.

Provision should be made for closers to be fitted to internal entrance doors to suit individual needs.

4.1.10 Entry phone systems should be carefully selected for:

- ease of location of appropriate dwelling on panel in multi-storey developments

- ease of operating call buttons

- response/information in both visual, eg LED (light-emitting diodes), and audible forms.

4.1.11 A pull handle fitted horizontally can assist in closing a door from behind.

Figure 4.1.11

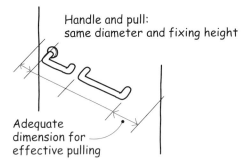

Handle and pull:
same diameter and fixing height

Adequate
dimension for
effective pulling

4.1.12 Entrances which are well lit externally are convenient as well as secure and lighting should illuminate all the key elements: door number, postal plate, bell, entry phone point, key and handle. Where lighting is not controlled as a common facility, PIR detectors should be considered in addition to switching. Door numbers and signing to common entrances should follow good practice given in the *Sign design guide* by Parker and Fraser (see Bibliography).

Figure 4.1.12

4.1.13 Lighting controls should be convenient in relation to the door, eg external light switching and two-way switching of hall and passage.

4.2   Requirements

4.2.1   Door

Provide effective clear width of at least 800 mm.

4.2.2   Approach space

Provide space beside leading edge of door, 200 mm minimum for a door opening away from the wheelchair user, 300 mm minimum for a door opening towards them, extending 1800 mm from face of door.

Figure 4.2.2

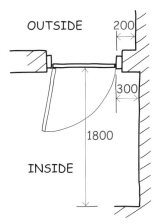

4.2.3 Threshold

Provide weathertight accessible detail to meet performance standard set out in *Design considerations* (4.1).

4.2.4 Lock

Provide secure locking, with key or locking control at 800–900 mm high.

Ensure easy to grip pull or handle.

Figure 4.2.4/4.2.7

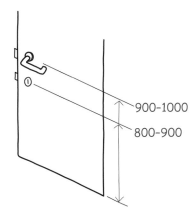

4.2.5 Opener

Install where opening pressure exceeds 20 newtons. Elsewhere make provision for installation of a remote-controlled door opener as an adaptation.

4.2.6 Lock, common external door

Where fitted with closer, provide remote-controlled door opener.

4.2.7 Lever, pull handles

Select for ease of use and good grip and to contrast with door. Fit at 900–1000 mm high.

Figure 4.2.7

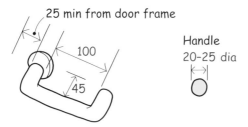

25 min from door frame

100

45

Handle
20-25 dia

4.2.8 Entry phone

Where installed, site 300 mm clear of any internal corner and at 1000 mm to buzzer.

Figure 4.2.8

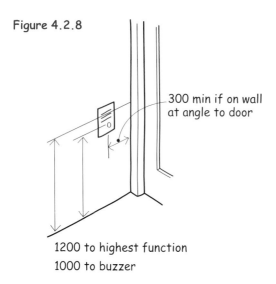

300 min if on wall
at angle to door

1200 to highest function
1000 to buzzer

4.2.9 Bell

Where no entry phone system provided, fit at 800–900 mm high with large vandal-proof operating surface, sited 300 mm clear of any internal corner. Contrast with door.

4.2.10 External light

Install local lighting which illuminates all external elements used by occupants, visitors and callers, and operates by PIR detectors, where not managed as a common facility.

4.2.11 Lighting

Provide accessible lighting controls in relation to door and external and internal routes.

4.2.12 Pull

Make provision to fit added closing pull on outer face of inward-opening door at 900–1000 mm high.

Figure 4.2.12

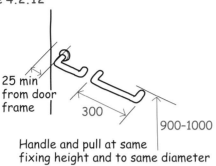

25 min from door frame

300

900-1000

Handle and pull at same fixing height and to same diameter

4.3 Recommendations

4.3.1 Door

Provide clear effective width of 900 mm.

4.3.2 Approach space

Provide 550 mm space beside lock edge of door opening towards wheelchair user.

Figure 4.3.2

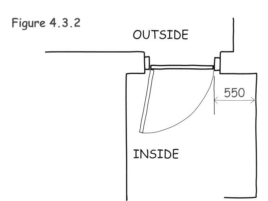

OUTSIDE

550

INSIDE

4.3.3 Entry phone

Where installed provide visual display to supplement audible response.

4.3.4 Number

Provide large raised numeral, contrast with background.

Figure 4.3.4

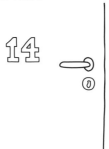

4.3.5 Lighting

Provide PIR detectors to entrance-light fitting in addition to switching.

Profile of a wheelchair user 6

A young woman, an amputee with spina bifida, lives alone with two young children. She has care support but is strongly independent. Her housing association purpose-designed, two-storey house has an integral lift which she is happy to use. The lift has a phone that she can use in an emergency. A neighbour has a key to the lift. She can use the stairs unaided (on her bottom). She has a manual wheelchair but cannot afford an electric wheelchair or a car.

In the living room, she worries about the security of the double external doors. The room's irregular shape makes it difficult to furnish. The narrower space, furthest from the kitchen, could be used as a dining area. She dislikes the lack of separation between the living area and the kitchen.

In the kitchen, the woman is happy with the adjustable worktop at 835 mm, and the side-hung door to the oven, with its drawer shelf underneath. The kitchen has a washing machine/drier, but there is no space for a separate drier. Stacking a drier would make it inaccessible. She can reach essential storage spaces in the kitchen, although airing cupboard shelves in the house are not within easy reach. The bathroom's textured floor is difficult to keep clean. The woman intends to carpet the upstairs bathroom.

With young children in the house, she is concerned about the safety of the front hob controls. Sometimes her knuckles touch hot radiator surfaces when pushing both chair and pushchair. She is generally unhappy about her inability to change light bulbs or take down curtains.

5 Entering and leaving; dealing with callers

Principle

Ensure that the wheelchair user in their own home can:

- enter, manoeuvre outdoor chair to transfer to indoor chair, and reverse the process when leaving
- leave outdoor or indoor chair on charge
- approach door to receive deliveries, retrieve post, open door to visitors, manoeuvre and return into living areas
- respond to callers and visitors without going to the door

5.1 Design considerations

5.1.1 Where more than one chair is used, the procedure for transfer between chairs will vary with different users. Transfer and charging space can often be the same space. It may be an effective use of space to incorporate the charging, transfer and space required to approach the front door into one location. Either or both wheelchairs may be electric and a power source must be provided.

Figure 5.1.1

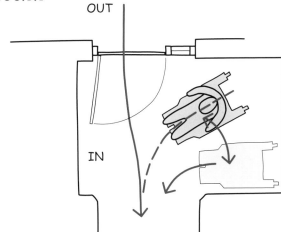

5.1.2   It should be possible for the wheelchair user to approach the door to collect post, respond to callers or open the door to let visitors in, then turn back into the home. Such turning space should be clear of all door fittings, including post collection fittings, when the door is closed.

Figure 5.1.2

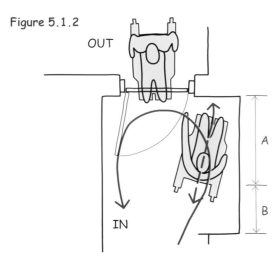

Unobstructed space to turn, A, and to manoeuvre, B

5.1.3   Visitors may well themselves be in wheelchairs and more space inside the entrance for visitor and host would be desirable.

5.1.4   It should be possible to collect post and papers without the need to reach to a low level. A collection fitting can be fitted to the door itself or within a side panel. Access to the door and its effective opening should not be compromised by such a fitting.

Figure 5.1.4

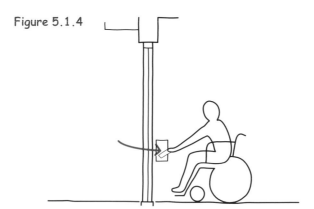

5.1.5   Some developers' policies may be to install an entry phone system at the outset, others to make provision for an intercom or a full entry phone system to be installed later. Hands-free fittings are more easily operated than the telephone type.

5.1.6 Passive security provision at an entrance can include glazing within the door, or to one side, door viewers or overlooking windows, eg from living room bays.

Figure 5.1.6

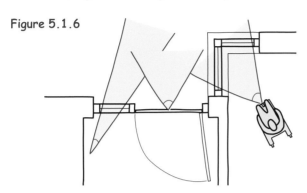

5.1.7 Entrance halls and spaces should have a good standard of natural light to reduce reliance on artificial lighting.

5.1.8 Most occupants will add matting to the floor finish to clean wheelchair wheels. Detailing should ensure that this will be practicable, ie that the door will clear the matting. The wheel circumference of the average self-propelling wheelchair is 1900 mm and to be effective matting may need to extend over the whole manoeuvring area.

5.1.9 Entrance lobbies are desirable since the outer entrance door may need to be open longer for manoeuvre and passage. However, they will not be practicable unless they provide sufficient space for a wheelchair user to clear one door swing before negotiating the other door.

5.2 Requirements

5.2.1 Transfer

Provide space within the house to manoeuvre wheelchair to transfer to a second chair, to store the first, and if necessary to leave it on charge, clear of circulation routes and the reqired approach to furniture and doors.

Figure 5.2.1

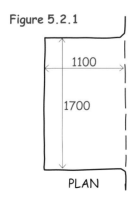

1100

1700

PLAN

5.2.2 Turning space

Provide space to manoeuvre and turn, 1500 × 1800 mm, clear of fittings and obstacles on a closed door.

Figure 5.2.2

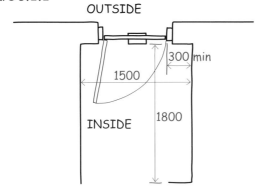

5.2.3 Post

Provide fitting to collect post while maintaining effective door opening width.

Figure 5.2.3

Letterbox fitting, operating height, on inside panel or door itself

5.2.4 Entry phone

Provide at common and individual entrance doors within flatted developments. At front entrance doors to individual dwellings make suitable provision for future installation, for example by providing blanked socket outlet, conduit and draw wires with operating points in living room and main bedroom.

5.2.5 Lobby

Where entrance lobby incorporated or provision made for added inner door, ensure adequate space to manoeuvre between doors as detailed in 5.2.2 and 7.2.7/7.2.8.

5.3　Recommendations

5.3.1　Enter and transfer

Provide space within the entrance area to manoeuvre wheelchair and transfer to a second chair, to store the other chair, or leave on charge outside the circulation area.

Figure 5.3.1

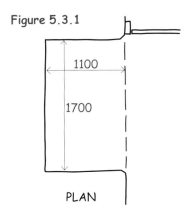

1100

1700

PLAN

5.3.2　Visitors

Provide space additional to 5.2.2 to approach and open door, receive visitors and allow them to enter, then turn to return into home.

Figure 5.3.2

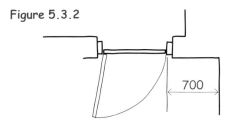

700

5.3.3　Callers

Provide for visual observation of callers.

Figure 5.3.3

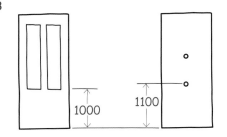

1000

1100

Glazed panels　　　Upper and lower viewers

5.3.4　Entry phone

Provide at front entrance doors to individual dwellings.

Install or make suitable provision for third operating point in kitchen (larger houses).

6 Negotiating the secondary door

Principle

Ensure direct connection to external spaces by an easily operated but secure door, as access to private or shared gardens and as escape in emergency. See also Section 2 *Using outdoor spaces* for consideration of balconies.

6.1 Design considerations

6.1.1 The door will open off circulation, kitchen or living spaces as determined by developer's policy and internal/external planning. Its siting should not generate routes across rooms which inhibit furnishings.

Figure 6.1.1

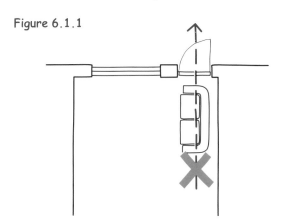

6.1.2 Where doors open off confined spaces such as passages or kitchens, there should be adequate space to approach and manoeuvre as set out in 4.1.3 and 4.1.4.

6.1.3 The door may be similar to the front entrance door, ie solid or semisolid, or it may be an integral part of the glazed areas.

Figure 6.1.3

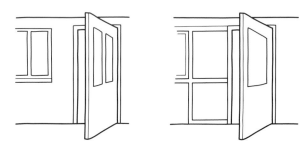

6.1.4 A single, adequate width door may be more appropriate for access than paired doors, in particular for compatibility with an added door opener and for security and operation. Sliding doors as currently available usually present barriers of threshold and operation which are insurmountable without an expensive recessed detail. They can also present problems of maintenance.

Figure 6.1.4

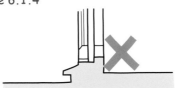

6.1.5 Doors may be outward opening to avoid loss of internal space but air or wind pressure may make them difficult to close and vulnerable to damage and appropriate fittings are needed. Inward-opening doors may be a more practical solution provided space allowance is made.

6.1.6 It should be possible to secure the door at all times but to allow easy exit by means of a handle or simple non-key device.

6.1.7 Lighting controls should be convenient in relation to the door, eg external light switch and two-way switching of room light. Route to outside garden may need to be lit by means of two-way switching if it leads to rear parking, storage or refuse.

Figure 6.1.7

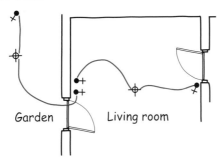

Garden   Living room

6.2 Requirements

6.2.1 Landing

Provide nominally level landing 1500 mm wide × 1500 mm deep
with 1200 mm clear of door swing.

Figure 6.2.1

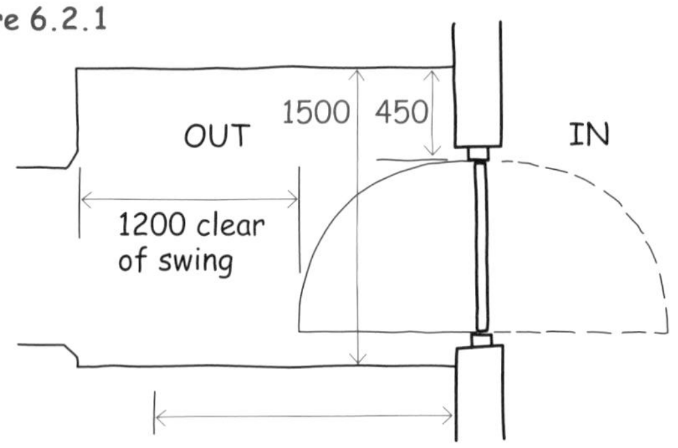

OUT   1500 450   IN

1200 clear
of swing

Can be reduced to 1500 when door opens in

6.2.2 Door

Provide effective clear width of 800 mm to single or main leaf.

6.2.3 Approach space

Ensure space to approach, manoeuvre and pass through door
on line.

6.2.4 Threshold

Provide weathertight, accessible detail.

Profile of a wheelchair user 7

An independent woman who is married with grown-up children. She has a
professional job and lives with her husband in their own three-bedroom adapted
bungalow.

She has two electric and two manual wheelchairs. Transferring, storing, and
re-charging these wheelchairs is difficult. She would prefer to have a large square
entrance hall with enough space for cleaning dirty wheels and for receiving visitors
who also use wheelchairs.

There is no entry phone for the main entrance, which would benefit from better
lighting. An infrared opener/closer would be useful.

The woman sometimes has problems using an electric wheelchair outside since it
has a tendency to skid on cambers. She would use buses if she had easier access to
them. Since shopping trips are not easy she relies heavily on her freezer.

The kitchen has a low sink (815 mm) with space underneath. There are low and
high worktops with base units. There is also a fitted ironing board.

The bathroom has a shower but no bath. She would prefer it to have a bath, a
separate WC and a bidet.

6.2.5 Lock

Provide to sole leaf, or where paired the main leaf, secure locking system offering:

- operation on latch
- security from outside when required
- simple operation from inside without key at all times
- key operation at 800–900 mm
- handles at 900–1000 mm high

Where applicable, provide to secondary leaf secure multi-locking (independent of main leaf) by single inside handle at 900–1000 mm high.

Figure 6.2.5

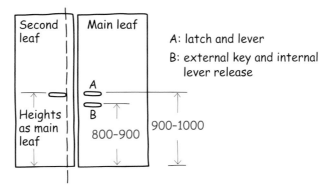

6.2.6 Stays

Provide to outward-opening door integral and adjustable fittings to prevent over opening or sudden closing in windy conditions.

6.2.7 Pull

Provide for addition of closing pull at 900–1000 mm high to suit occupant.

Figure 6.2.7

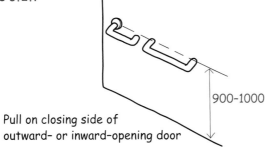

6.2.8 Lighting

Provide accessible lighting controls in relation to door and external and internal routes, with operation of external lights by PIR detectors.

7   Moving around inside; storing things

Principle

Ensure that wheelchair users can:

- conveniently manoeuvre, approach and negotiate all doors within circulation areas

- approach and use storage off circulation areas

7.1   Design considerations

7.1.1   The key determinant of adequate circulation is space to manoeuvre a wheelchair conveniently and without personal discomfort to hands, fingers, etc, or damage to building fabric. The flexibility to incorporate some internal circulation into living areas may be an effective use of space and a consideration in many cases, but best practice in larger family dwellings is to be able to access most rooms off circulation areas.

Figure 7.1.1

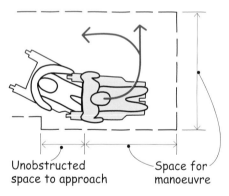

Unobstructed space to approach

Space for manoeuvre

Profile of a wheelchair user 8

An older man, a double amputee, lives alone in a one-bedroom, ground floor, adapted, local authority flat. He has care support. The flat has an entry phone, plus door opener with infrared handsets.

The man has two electric wheelchairs, one indoor and one outdoor. He uses space off the hall to park and re-charge his wheelchairs, and uses a hoist in the bedroom to transfer between them and the bed.

He needs a hoist in the bathroom for quick access to the WC and shower. There is no bath. He uses his knuckles to operate full-plate switches that are placed at 900 mm. Sockets are placed at 800 mm (not too low for him). He operates these with his elbow.

7.1.2   There are a number of critical wheelchair manoeuvres for which adequate space must be provided to suit a reasonable range of users:

- 90° and 180° turns within circulation routes

- 90° turns to approach and pass through doorways

- angled approaches to doors

- head-on approaches to doors opening towards or away from the wheelchair

Figure 7.1.2

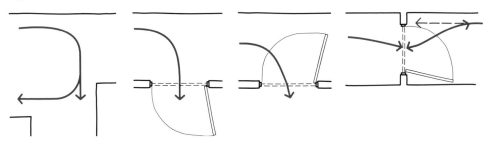

7.1.3   The minimum dimensions set out in the following paragraphs take account of those users who by virtue of their chairs, whether manual or electric, or their own more limited capabilities may require a reasonable degree of spatial tolerance in carrying out such manoeuvres. This is particularly so at doorways where the relationship of approach space and clear door opening width may be critical, as discussed below. The one-off dwelling may require a further increase in standards.

7.1.4   Hallways or circulation routes should have the following minimum clear widths:

- at least 900 mm for straight passage

- 1200 mm to allow a 90° turn into a doorway

- 1500 mm to allow a 180° turn

Clear means between finished wall surfaces or any projections (apart from skirtings) such as radiators or parked wheelchairs. Such projections should be kept clear of the full space needed for manoeuvre.

Figure 7.1.4

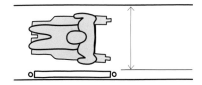

7.1.5 To allow turning at right angles there should be 1200 mm clear width in each direction, although 900 mm in one direction will be manageable by most.

Figure 7.1.5

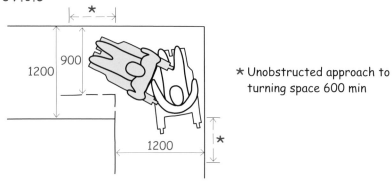

∗ Unobstructed approach to turning space 600 min

7.1.6 Outer angles within circulation areas could be splayed or rounded. Small splays will minimise damage and may be preferable for corner protection. Larger ones, including the skirting, will ease circulation and may allow passage widths to be reduced to 900 mm in each direction.

Figure 7.1.6

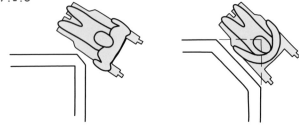

Profile of a wheelchair user 9

A young woman, severely physically disabled with very limited reach. She lives with care support in a purpose-designed, housing association, two-bedroom, single-storey dwelling.

The entrance has an entry phone and the front and back doors are fitted with an infrared door opener/closer.

In the kitchen, the 900-mm-high worktop fits over the tumble drier but a reduction of only 25 mm is available to accommodate the wheelchair's arms and joystick. The sink and taps need major adaptation for the woman to use them. Although most kitchen activities are carried out by carers, she would like the scope to contribute, and to manage small tasks.

In the bathroom, she is unable to reach the shower control on the back wall. Nor can she reach the bath's corner taps which are located against the end wall. She is also unable to open and close windows, but can operate all other switches.

7.1.7 A circulation layout which is square rather than linear may be more convenient and no less efficient in use of space. Such a layout would be determined by the doors opening off it rather than by the relationship between door opening and passage widths. The relationship between the entrance and the internal circulation may be significant in achieving this. Excessive hallways that create an institutional feel should be avoided.

Figure 7.1.7

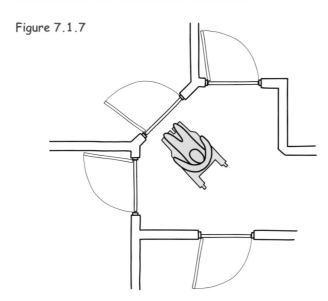

7.1.8 The provision of effective clear width to internal doors is critical. There is a relationship between the way in which a door is approached and negotiated and the effective clear width of the door. It may in practice be more economical to adopt a wider door standard throughout the house than to vary doorset widths. Ample clearance will help to minimise damage from wheelchairs and avoid the need for added protection.

7.1.9 Where practicable, doors should open further than 90°, but not more than 110°, both to ease turning and to ensure that handles do not affect the effective clear width. See also *Doors, basic criteria* on page 12.

Figure 7.1.9

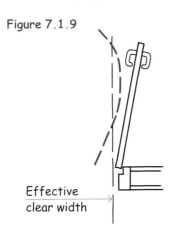

Effective clear width

7.1.10 An effective clear width of 775 mm may be followed for a door approached on line or opening out of a passage which is at least 1200 mm in clear opening width. In other contexts, as noted below, an increase in effective clear width may be advisable.

Figure 7.1.10

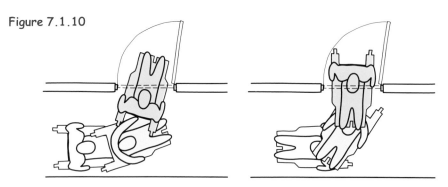

7.1.11 The approach to a door opening outwards into a passage results in the user passing through the doorway at an angle. A further increase in effective clear width is therefore desirable to avoid damage and inconvenience.

Figure 7.1.11

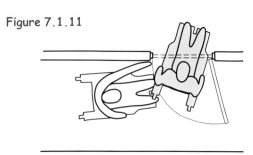

7.1.12 Similar considerations apply for leaving a bathroom, where the fixtures and fittings may preclude on-line passage through the doorway.

7.1.13 The head-on approach to a door opening towards the user usually necessitates an awkward manoeuvre. The familiar 300 mm space beside the latch edge of the door facilitates the manoeuvre of approaching to release the door then reversing in the wheelchair while opening the door. Where possible an increase to 550 mm will greatly ease this manoeuvre. There should be clear unobstructed space to reverse clear of the door swing.

Figure 7.1.13/7.1.14

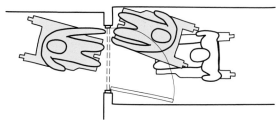

7.1.14 The head-on approach to a door opening away from the user is facilitated by a minimum 200 mm space beside the latch edge of the door.

7.1.15 The approach to a door which requires a 90° turn at the end of a passage is facilitated if an outward-opening door opens beyond 90° or if there is space beside the edge of an inward-opening door.

Figure 7.1.15

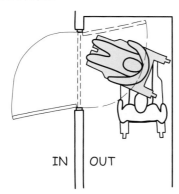

IN | OUT

7.1.16 The preceding considerations refer primarily to conventionally hinged doors. Doors are usually hinged, but sliding, double-swing or reduced-swing doors may be more appropriate in some situations, for example where doors would otherwise open out into passages, provided a clear opening width can be achieved and the appropriate approach allowed for. For selection of other door types see Section 13 *Operating internal doors*.

7.1.17 The approach to a sliding door is usually at an angle for easier operation, necessitating space beyond the latch edge and an increase in effective clear width.

Figure 7.1.17

7.1.18 The best protection for doors, walls and linings is provision of enough space for ease of manoeuvre. Added protection such as kick plates or angles, which can lead to an institutional appearance, should only be as an adaptation to suit the demands of the user and chair which are in excess of the standard range.

Figure 7.1.18

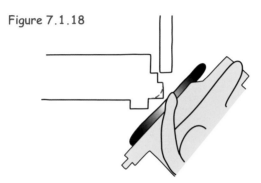

7.1.19 A good level of natural daylight reduces reliance on artificial lighting and the need to change light bulbs more frequently.

7.1.20 Storage should be adequate for the size of the house with additional provision for equipment specific to a wheelchair user. It should meet the need to be able to put away and take out bulky household equipment such as cleaners. Depth of storage and clear opening width of doors should permit maximum access to space and shelving. For a wheelchair user, wide but shallow storage approached sideways is a more efficient use of space than deep narrow storage where it may be difficult to reach items stored at the back.

Figure 7.1.20

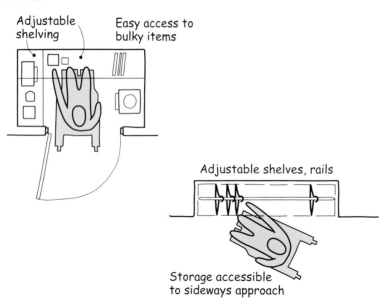

7.2   Requirements

7.2.1   Straight passages

Ensure that passage widths or approaches where no turning or door approach is required are no less than 900 mm clear of all obstructions (except skirtings).

Figure 7.2.1

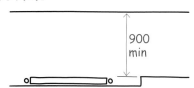

900 min

7.2.2   Head-on approach to doors in passages

Ensure space beside latch edge of door, minimum 200 mm on push side and minimum 300 mm on pull side.

7.2.3   Turning 90°

Ensure that passage widths or approaches to turn through 90° are no less than 1200 mm clear of all obstructions (except skirtings) for extent of manoeuvring space.

7.2.4   Turning 180°

Ensure that passage widths or approaches to turn through 180° are no less than 1500 mm clear of all obstructions (except skirtings) for extent of manoeuvring space.

Figure 7.2.4

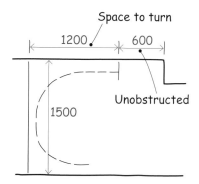

Space to turn

1200   600

Unobstructed

1500

7.2.5 Right angles

Ensure at right angles that passage width clear of all obstructions (except skirtings) for the extent of the turn is no less than 1200 mm width in one direction, and 900 mm in the other; or 900 mm in each direction in combination with angle splayed by 300 mm.

Figure 7.2.5

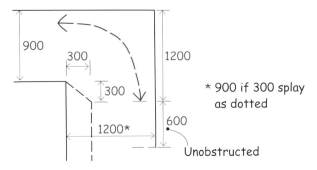

* 900 if 300 splay as dotted

7.2.6 Effective clear widths for doors

Ensure 775 mm minimum effective clear width. Increase where approach is at an angle.

Figure 7.2.6

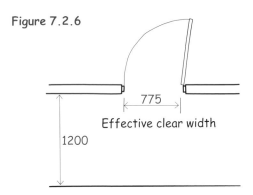

Effective clear width

7.2.7 Space to approach doors

Provide space beside latch edge of doors:

- 300 mm minimum for doors opening towards user
- 200 mm minimum for doors opening away

Figure 7.2.7

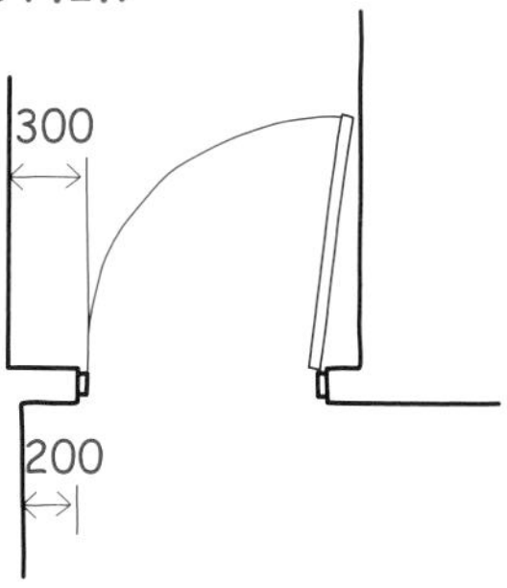

7.2.8 Doors at angles

Provide space to turn between doors at an angle to each other.

Figure 7.2.8

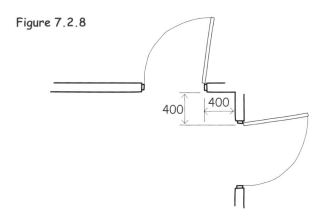

7.2.9 Sliding doors

Provide space beyond doorway at latch side for sideways approach and operation.

Figure 7.2.9

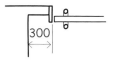

From return wall

7.2.10 Storage

Ensure that depth and width of storage space, in combination with any shelving layout, provides optimum access to space and to stored items. Ensure that opening width of doors suits angled or head-on approach.

7.3 Recommendations

7.3.1 Circulation

Where practicable, consideration should be given to providing a centralised layout where doors open off a core space rather than passages, or to reducing circulation requirements by incorporating some circulation within rooms.

7.3.2 Passages

Ensure that passage widths or approaches where no turning or door approach is required, are no less than 1000 mm clear of all obstructions (except skirtings).

Figure 7.3.2

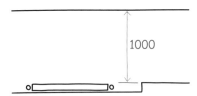

7.3.3 Angles

Provide splays to all outer angles, including skirtings.

Figure 7.3.3

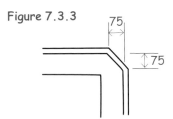

7.3.4 Clear door opening width

Ensure 800 mm minimum effective clear width to all doors.

7.3.5 Space to approach doors

Provide 550 mm minimum space beside latch edge of door opening towards user.

Figure 7.3.5

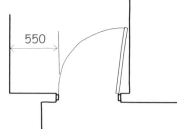

7.3.6 Doors at end of passage

Ensure that outward-opening door opens beyond 90° and that there is 200 mm beside edge of door opening inwards.

Figure 7.3.6

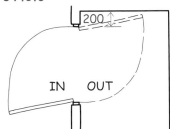

7.3.7 Opposing doors

Ensure that doors on opposing sides of a passage are directly opposite, preferably in mirror image.

Figure 7.3.7

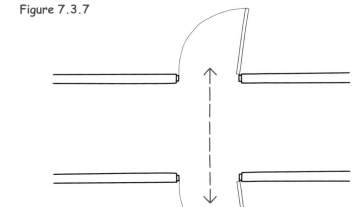

Profile of a wheelchair user 10

A young woman, severely disabled after a car crash, is about to move to her own adapted and extended bungalow. She intends to have maximum control over her home activities and environment but will rely on live-in carers, one of whom is her mother.

The woman has a manual wheelchair but intends to obtain an electric one to increase her mobility. She has an adapted VW Caravelle with side access which is too large for her needs. She has very limited physical ability and reach; her vertical reach from a sideways position is between 700 and 1000 mm.

She is having a customised system installed to give her independent control of external doors, key internal doors, windows, curtains, equipment, telephone, etc. She has requested a shower rather than a specialist bath (for independence), an automatic wash/dry toilet, and a hairdresser's basin in addition to a conventional one.

8 Moving between levels within the dwelling

Principle

Where dwellings are designed with more than one floor level, ensure that there is provision for independent movement between floor levels in a wheelchair without the need to transfer, without compromise of circulation or living space, and with all rooms remaining accessible.

8.1 Design considerations

8.1.1 The principle implies a lift connecting circulation at each level as an integral element rather than an adaptation.

8.1.2 Space requirements for a through-floor lift are considerable, embracing lift shaft and specific circulation in addition to stairs. Developers will need to balance initial and ongoing costs against other economic factors.

8.1.3 Domestic lift installations may be self supporting between two floors or incorporate tracking mounted on end or flank walls. A typical clear internal car size is 1085 × 725 mm for a front-entry lift but research into manufacturers' products is advisable to ensure at the outset adequate shaft and floor-opening sizes.

8.1.4 The space to approach a lift will be similar to that needed to approach an outward-opening door, but the restricted opening width of car doors may require a head-on approach and an increase in approach width beyond 1200 mm unless the lift door is remote controlled. See also Section 7 *Moving around inside; storing things.*

Figure 8.1.4

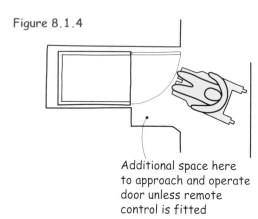

Additional space here
to approach and operate
door unless remote
control is fitted

8.1.5 Safety considerations imply emergency call, eg a telephone or alarm to reliable support outside the home, manual override and a return to lower floor facility. There should be a full range of integral features to comply with current standards and best practice.

8.2 Requirements

8.2.1 Lift

Provide lift for independent use by a wheelchair user to BS 6440 which connects floors and is accessed off circulation spaces at each floor level.

8.2.2 Installation

Include a full range of safety and security features such as:

- integral and fire-resistant ceiling and floor traps
- sensitive edges and surfaces
- manual override and return to entrance floor
- lift-mounted telephone and/or alarm

8.2.3 Circulation

Provide adequate circulation space at each level to manoeuvre, call lift, approach and open its door and use it.

Profile of a wheelchair user 11

A young man lives alone in a purpose-designed, two-bedroom, housing association, single-storey house. He shares ownership of the house. He has a manual wheelchair.

The entrance has an entry phone and external lights on sensors. He is happy with this arrangement but would also like an infrared door release. There is a 1400-mm-wide passage leading from the front door. All doors are wide enough at 825 mm except for the store cupboard door which at 725 mm is too narrow for access.

The man is unable to reach the kitchen windows. The worktop is set at 800 mm, with raised pipes underneath, although the stop tap is inaccessible.

In the shower-only bathroom there is room to fit a hoist from the bedroom with en-suite door, if required. The bedroom has a single bed; it would be inaccessible with a double.

9 Using living spaces

Principle

Ensure that a room can accommodate the usual range of furniture with space for a wheelchair-using member of the household to circulate and transfer from wheelchair to seating. See also *Layout of rooms, basic criteria* on page 14.

9.1 Design considerations

9.1.1 The relationship between kitchen and living spaces should allow individual households to balance ease of movement with preferred living and dining arrangements according to household size and preferences. In smaller houses provision of partial screening of the kitchen within an open arrangement, which could be completed by the subsequent addition of a sliding door, might be considered.

Figure 9.1.1/9.1.2

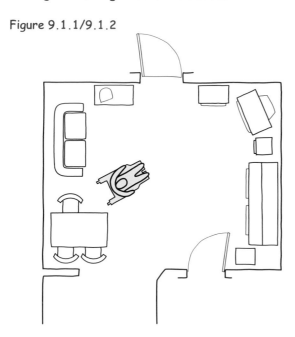

9.1.2 Some occupants may opt for a perimeter layout of furniture to maximise manoeuvring space. Layouts should avoid constraints on furniture layout and movement such as window positions, doors entering in a corner, rooms or spaces narrower than 3000 mm.

Figure 9.1.2

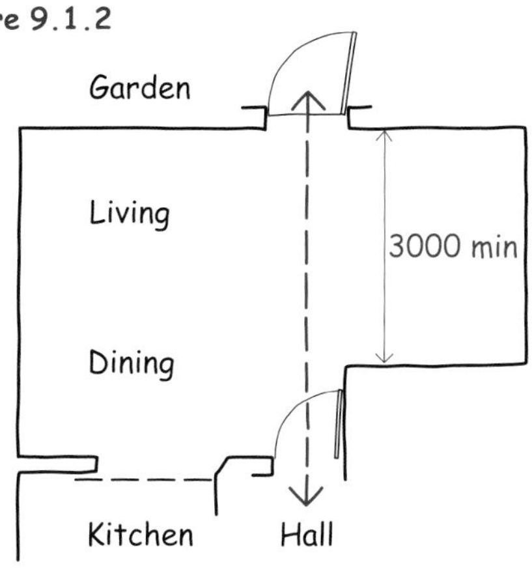

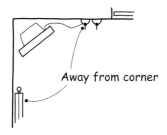

Garden

Living

3000 min

Dining

Kitchen   Hall

9.1.3 Spaces for study, computer and internet use should be considered.

9.1.4 Radiator and socket outlet positions should be considered carefully in order not to restrict disposition of furniture or audio-visual equipment. See also Section 15 *Controlling services*.

Figure 9.1.4

Away from corner

Profile of a wheelchair user 12

A young man with very limited physical ability lives alone in a one-bedroom, adapted, ground floor, local authority flat. He is largely independent, has support from friends and assistance from a carer. He has two electric wheelchairs, and transfers independently and laterally in the living room where there is adequate space.

The flat is only manageable because it has sliding doors, and if he does not use footrests on his wheelchair. The man needs to approach the front door sideways in order to support his key-holding arm. The canopy is important. The full plates light switches are specially set at 800 mm – just above the wheelchair arm rests (715 mm). This reduces to a minimum his need to raise his arm and wrist.

The flat has an entry phone and remote control of windows is being arranged. The bathroom contains a shower but no bath. Space for manoeuvring is very limited.

9.2 Requirements

9.2.1 Room layout

Provide space for furniture and for a wheelchair user to approach it, circulate, transfer to seating, and approach and operate doors, windows, equipment and controls.

9.2.2 Radiators

Ensure that their positioning does not inhibit a reasonable layout in terms of wheelchair movement.

9.2.3 Sockets

Ensure that sockets are not sited within 750 mm of an internal room angle.

Figure 9.2.3

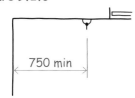

750 min

9.3 Recommendations

9.3.1 Where possible avoid doors opening into corners of rooms.

Profile of a wheelchair user 13

A professional man lives on a small estate in a purpose-built wheelchair user bungalow with his wife and the youngest of three grown-up daughters. He transfers to a chair at home. He uses a seat stair-lift to reach the upper floor of his adapted offices where he then transfers to a chair.

Until recently the man drove independently, transferring to the driver's seat from the passenger side, and pulling the folded wheelchair in after him. The capability to transfer under cover is important to him because it keeps the seat dry in the rain. The small size of the integral garage prevents him from buying a larger and more comfortable vehicle since it would have to be parked unprotected in the street. It would also have to be accessed via an excessively steep path. He also finds rough textured paving painful.

The house is quite tightly planned but manageable by virtue of its central hall (2400 mm × 1600/1800 mm and 826-mm-wide doors, and 900 mm doorsets).The man has a number of minor complaints. The window sills are too high for comfortable viewing out. From ground to sill is 810 mm, it is 870 mm to the glass, but the transom height is too low at 1100–1200 mm. The man is unable to change light bulbs. The third bedroom, which is now spare, has an 800 mm doorset. It could be a useful office or photographic darkroom.

10 Using the kitchen

Principle

Ensure ease of approach to and use, from a wheelchair, of the sink, worktops, equipment, all appliances and controls and all storage essential to kitchen operations.

10.1 Design considerations

10.1.1 The primary objective should be to allow the wheelchair user to operate within their own kitchen. However, it should be acknowledged that during the life of the dwelling, disabilities may vary between successive occupiers and the wheelchair user may not always be the main user.

10.1.2 The key to good kitchen design for wheelchair users is a layout which provides effective and appropriate space. The layout should avoid compromising working areas with cross routes. Except in smaller dwellings, the layout should allow for concurrent use by more than one person. There should be good-quality well considered daylighting. See also Section 14 *Operating windows*. Artificial lighting should combine general lighting with well positioned shadow-free task lighting locally or two-way switching.

Figure 10.1.2

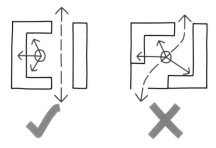

10.1.3 Established good practice in kitchen design such as sequences, continuity and travel distances, should be observed in conjunction with the following specific considerations.

10.1.4 A wheelchair user needs space under the worktop at hob, sink and other critical points in order to operate equipment, reach controls, and carry out essential activities including access to related storage. Space to enable this should be not less than 600 mm wide.

Figure 10.1.4

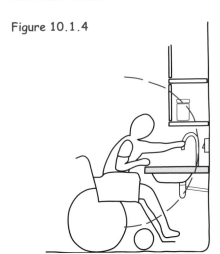

10.1.5 The layout should enable a wheelchair user to do this without the need for excessive manoeuvring or positioning. The layout should therefore maximise the range of operations possible from one wheelchair position. An L-shaped arrangement may facilitate this, as may approach space wider than the minimum 600 mm width.

Figure 10.1.5

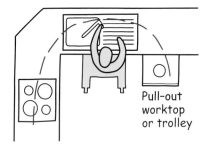

Pull-out worktop or trolley

Profile of a wheelchair user 14

A single young woman lives in a two-bedroom ground floor flat in a housing association block where she has had minor modifications made. She is largely independent but has some assistance from her mother. She has one manual wheelchair and doesn't drive a car.

The outer door and front door can be opened using remote control. She dislikes the lack of an alternative external door.

The kitchen is unadapted but the woman doesn't cook much. The bathroom has a shower only. In the bedroom, she prefers a double bed for comfort.

10.1.6 Housings for built-in appliances or appliances which extend above worktop level should be carefully positioned for:

- ease of approach and operation
- effective relationship to essential activities
- avoidance of disrupted sequences or operations

Figure 10.1.6

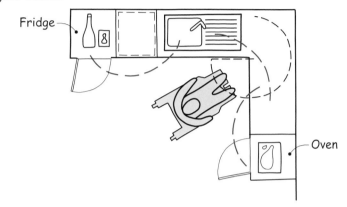

10.1.7 The height of worktops may need to be adjusted within the life of the kitchen fittings. Whether adjustability is facilitated by built-in provision from the outset or whether a compromise height is determined initially will depend on how the following factors are weighed against initial or subsequent costs.

Flexibility in socket heights and plumbing connections should be considered. See 10.1.16 and 10.1.24.

Figure 10.1.7

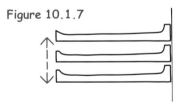

10.1.8 Where it is known that the main user is a wheelchair user, it should be possible to suit his or her needs closely.

10.1.9 An incoming wheelchair user may, from lack of experience, need time to experiment before determining the optimum height for them, or they may require a different height from a previous occupant.

10.1.10 Where the wheelchair user is not the main user of the kitchen, there may be a need to determine a compromise between different users – wheelchair user and ambulant.

10.1.11 Adjustable brackets (excluding angle brackets) may in any case be an effective and economic means of achieving clear space below worktops.

Figure 10.1.11

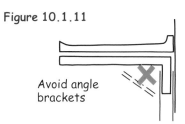

Avoid angle brackets

10.1.12 It may be helpful, particularly where there is shared use, to be able to adjust different sections, for example worktop with sink, worktop with hob.

Figure 10.1.12

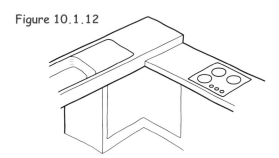

10.1.13 Where worktops are intended to be adjustable simply, the worktop-to-wall detail should ensure that wall tiling is not disturbed. The depth of the latter should suit the range of heights.

Figure 10.1.13

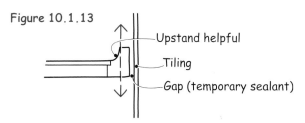

Upstand helpful

Tiling

Gap (temporary sealant)

10.1.14 At inner angles of worktops it can be difficult to reach items or switches and sockets. Switches and sockets should be positioned away from the corner. A splayed section of worktop wide enough for wheelchair approach would be helpful.

Figure 10.1.14

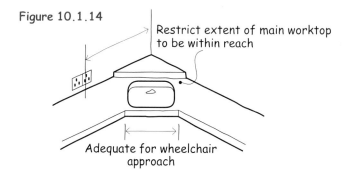

Restrict extent of main worktop to be within reach

Adequate for wheelchair approach

10.1.15 Pull-out worktops are useful both in providing a lower surface and in maximising operations from one position.

Figure 10.1.15

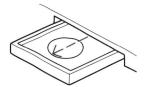

10.1.16 Sinks should be shallow for optimum approach except that a minimum 150 mm bowl depth is preferred for family use. Wastes should be to the rear of the bowl and the underside should be well insulated to protect the wheelchair user when approaching. Preparation, rinsing and draining can be facilitated by means of a second bowl or separate drainer, or both. Inserts such as baskets or chopping boards can maximise effective working area. Plumbing should be flexible or otherwise easily modified to suit adjustable sink heights.

Figure 10.1.16

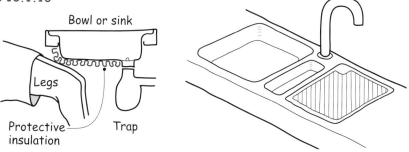

10.1.17 Taps should permit easy control of water temperature, flow and direction. This suggests a single-hand temperature and volume control of the mixer and a swivel arm of sufficient height and radius to permit filling pots, etc, on the surface adjacent to the sink bowl. Provision for separating taps from swivel arm, eg by mounting to side of sink or on fascia, could be helpful to suit limited reach, possibly as an adaptation.

Figure 10.1.17

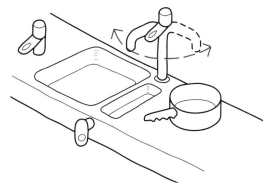

10.1.18 Space under the kitchen sink is potentially untidy. It may be possible to site the sink out of view in more open-plan layouts. Care should be taken over plumbing with exposure to view in mind and to avoid injury from foot contact with hot surfaces and pipes.

10.1.19 Storage presents problems because wheelchair users need to approach under worktops and have limited reach. In general terms storage, reachable from a wheelchair, needs to be within the height range 300–1500 mm and not deeper than 300 mm whether above or below the worktop. In consequence a small proportion of storage such as the top shelves of wall-mounted units may be out of the reach of the wheelchair user. Some flexibility of location and operation is desirable.

10.1.20 Possible ways of maximising flexibility and access from a wheelchair could include:

- storage trolleys
- large carousels
- pull-out units to deep spaces
- cantilevered shelf brackets to wall cupboards
- doors opening beyond 90°

Trolleys can also be useful in transferring items, for example from oven to worktop or table.

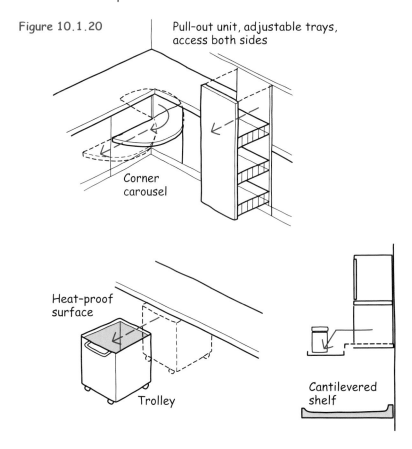

Figure 10.1.20

Pull-out unit, adjustable trays, access both sides

Corner carousel

Heat-proof surface

Trolley

Cantilevered shelf

10.1.21 Handles to drawers and cupboards should be robust and easy to grip and should contrast with surfaces. Push/spring type catches are available.

Figure 10.1.21

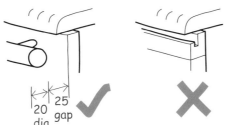

10.1.22 Space and services should be provided for a range of appliances such as hob, oven, fridge, freezer, washing machine, drier, dishwasher.

10.1.23 Where it is possible or intended to reflect individual wheelchair user needs, the user should be able to determine and select appliances appropriate to their needs, preferences or capabilities. Preferences could, for instance, be for microwave rather than conventional oven. A side-hung rather than drop-down oven door may be preferable for safety and ease of transfer of food and utensils. Mobility and strength needed to operate controls and doors of some appliances may be significant.

Hobs with flat surfaces may assist transfer of pans, etc, but controls to the rear may be inaccessible and dangerous. Both hob undersides and drop-down oven doors should be well insulated where there is approach space underneath. Ovens should not be sited under hobs.

Figure 10.1.23

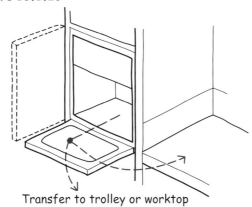

Transfer to trolley or worktop

10.1.24 Controls and switches are not easy to resolve satisfactorily for a reasonable range of user reach. Sockets for worktop equipment can be switched laboratory-type mounted on adjustable worktops, or unswitched wall or worktop mounted and linked to remote switches positioned on worktop fascia.

Appliances are usually connected to socket outlets at lower level with remote switches similarly mounted on fascia or conveniently wall mounted. A clear labelling or other identification system should be incorporated. See also Section 15 *Controlling services*.

Figure 10.1.24

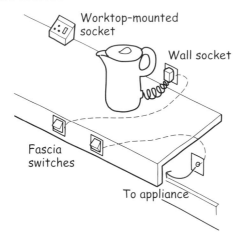

10.1.25 Reduction in provision of base units may increase floor areas to be cleaned. Plinths below corner carousel units or below equipment such as a washing machine will reduce such areas.

Figure 10.1.25

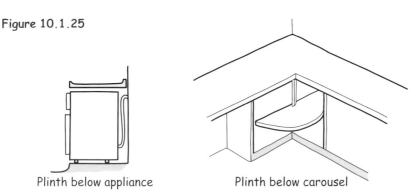

Plinth below appliance   Plinth below carousel

10.1.26 Other utility activities, such as ironing or disposing of refuse, need to be resolved within the kitchen or elsewhere within the dwelling to facilitate wheelchair use. An adjustable wall-mounted ironing board could be fitted as an adaptation.

10.2 Requirements

10.2.1 Layout

Take account of foregoing considerations and lay out the kitchen to provide a practical working kitchen for a wheelchair user. Ensure clear manoeuvring space not less than 1800 × 1500 mm. Ensure wherever practicable, that windows are positioned for ease of control and cleaning.

10.2.2 Worktops

Provide a length of 600 mm deep worktop suitable for a wheelchair user (with clear knee space below) and appropriate to the size of dwelling as set out in the National Housing Federation's *Standards and quality in development: a good practice guide* (see Bibliography).

10.2.3 Sink

Provide integral shallow sink and drainer to maximise height adjustability, with insulated bowl, suitable for a wheelchair user (with clear knee space below), and accessible, easily manipulated mixer tap with swivel arm extending over drainer or worktop.

10.2.4 Storage

Provide storage appropriate to the size of dwelling as set out in the National Housing Federation's *Standards and quality in development: a good practice guide*, the major proportion of which is in a position and format usable from a wheelchair. See also *Using a wheelchair* on page 10.

10.2.5 Controls and lighting

Provide all electrical controls, including sockets, within reach to suit adjustable worktops. Provide remote and labelled switches for appliances and equipment. Combine general lighting with well positioned task lighting.

Figure 10.2.5

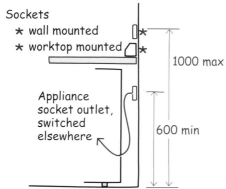

Sockets
  ★ wall mounted
  ★ worktop mounted

1000 max

Appliance socket outlet, switched elsewhere

600 min

10.2.6 Appliances

Provide and install hob and built-in oven. The hob should have a knee space below and the oven should be at an accessible height, adjustable to suit the user. Provide spaces in addition to hob and oven for three appliances/white goods such as fridge, freezer, washing machine, drier, dishwasher, with electrical and water services. Provide four such spaces in larger dwellings accommodating five or more persons.

10.2.7 Refuse

Provide suitable internal refuse arrangements, such as sealable sacks, manageable from wheelchair.

10.3 Recommendations

10.3.1 Worktops

Make provision for worktops, including hobs and sinks, to be adjustable in height in the range 750 to 910 mm.

Figure 10.3.1

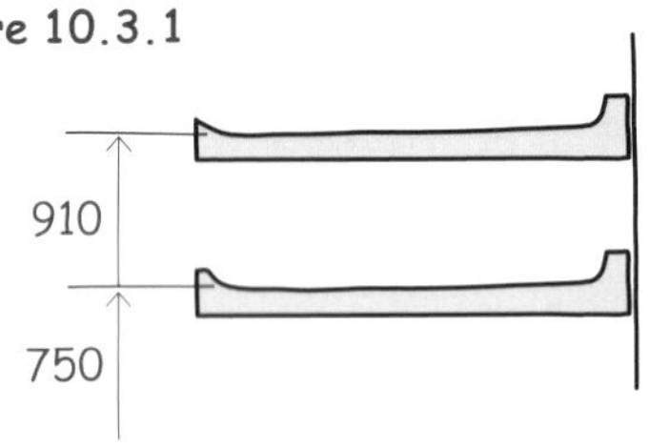

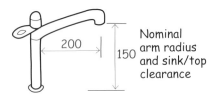

10.3.2 Taps

Provide easily controlled mixer with swivel arm extending over drainer or worktop.

Figure 10.3.2

10.3.3 Storage

Maximise storage accessible to wheelchair user by means of trolleys, carousels, pull-out baskets, shallow wall cupboards.

10.3.4 Appliances

Provide spaces for four appliances/white goods in addition to hob and oven with electrical and water services. Provide five such spaces in larger dwellings accommodating five or more persons.

11 Using the bathroom

Principle

In housing design terms the design of the bathroom is the key to enabling independence and dignity for disabled people. The ability to manage most if not all toileting and bathing functions without assistance is highly desirable and is the foundation of 'independent living'.

Ensure that there is scope for independent approach and safe transfer to all bathroom fittings, and for independent use of them. Ensure choice between shower, bath or both. Fulfilling bathroom functions privately is an essential part of independent living for wheelchair users. Well considered daylighting is desirable. See Section 14 *Using windows.*

11.1 Design considerations

11.1.1 Provision should suit size of household and take account of extra time a wheelchair user may take over bathroom activities.

11.1.2 The basic provision should be a fully accessible single bathroom containing a shower or bath, WC and basin. Where there is one bathroom a fully accessible and installed shower should be provided at the outset, with provision made for a bath in place of the shower area. Where any additional WC room is provided it should also be fully accessible and include a basin. The possibility of increasing the side transfer space within the latter to enable a shower to be installed should be considered. Any provision for a future shower installation to bathroom or second WC should include a shower outlet and gully flush with suitable flooring such as PVC laid to falls with coving at walls and conduiting for any future electrical supply. See 11.1.9.

11.1.3 Individuals may have preferences for right hand or left hand side transfer options to the WC. There should be provision for bathroom and WC layouts to be handed to suit individual capabilities. This can be done by ensuring that dwelling layouts within a development are handed and that the second WC within a dwelling is handed the opposite way to that of the bathroom.

11.1.4 An effective bathroom layout should provide appropriate space for general manoeuvre and to approach and use specific fittings taking account of the bath or shower provision. Individual users will have their own ways of approaching and transferring to WC, bath and shower to suit their capabilities and the relationship between them will be important.

11.1.5 The layout illustrated is a guide to such a layout and the following key factors should be noted:

- a minimum 1500 × 1500 mm manoeuvring space should be provided clear of all fittings; footrest space under fittings will maximise this

- the siting of WC should allow space clear of any door swing, be adjacent to a flank wall for firm support provision and allow for a full range of transfer methods – head-on, at an angle, sideways. Care should be taken with ducting and back-to-wall WC pans to ensure that the projection necessary for sideways transfer is maintained

- adequate space should be provided for a full length, 1700 mm, bath with, desirably, provision for an end transfer seat

- the space between bath or shower area and WC should allow access to bath taps, transfer from wheelchair to shower seat as well as to WC, and for use of shower chair within shower and over WC

- the basin should be sited clear of the frontal approach to the WC (see 11.1.11)

- there should be provision for direct access from an adjoining bedroom (see 11.1.7)

Figure 11.1.5/11.1.7

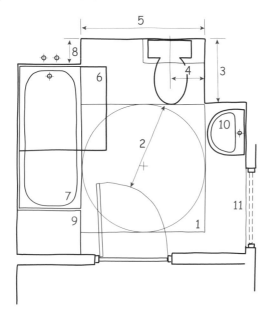

1 1500 x 1500 square manoeuvring space clear of all fittings, 1500 dia turning circle superimposed

2 1100 min between inward-opening door swing and WC

3 750 projection of WC for lateral transfer from wheelchair, whether freestanding WC suite or back to wall pan and ducting

4 450 centreline of WC to flank wall

5 1500 min to provide lateral transfer space

6 1000 x 1000 shower activity area
or
7 1700 x 700 bath

8 300 nominal to give access to bath taps and allow access/transfer to shower area

9 End transfer space with cupboard below

10 Basin or vanity top clear of frontal approach to WC

11 Knock–out panel for connecting door from bedroom detailed to suit hoist track installation

Figure 11.1.5

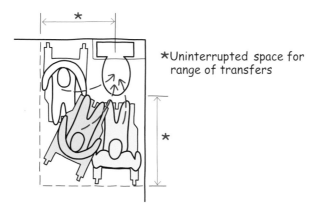

*Uninterrupted space for
 range of transfers

11.1.6 Other factors should be considered.

- The door into the bathroom should be carefully positioned to optimise manoeuvre between fittings and to ensure also that it can be approached easily on leaving. If the door opens into the bathroom, manoeuvring space must be provided beyond its swing or operating space. There should be provision for opening an inward-opening door from outside in emergency. See Section 13 *Operating internal doors*.

- A shower area (whether it is fitted-out initially or provision is made) which doubles as transfer space, eg between WC and bath, is an efficient use of space but has inherent defects. The controls are sited on the rear wall making it difficult for some to use, and effective enclosure is difficult. Where a fully accessible second WC is provided it could be a better solution to incorporate shower provision within it. Such a solution would be appropriate where both bath and properly accessible shower are required.

- Siting the basin for use from a seated position on the WC is not as critical as other factors although some users regard it as very important. It could be incorporated in the second WC where it would also benefit visitors in wheelchairs. More detailed guidance is given in BS 8300.

Figure 11.1.6

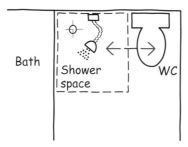

Bath

Shower space

WC

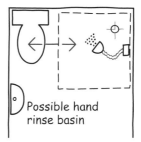

Possible hand rinse basin

11.1.7 Direct access from the main bedroom into the bathroom should be made possible whether a door is incorporated from the outset or its future provision is allowed for by means of a knock-out panel. Entry into the bathroom in proximity to the main door may not significantly increase the bathroom area if at all and enable the layout to remain unaltered.

The potential need for ceiling-mounted hoist tracking between bedroom and bathroom should be addressed and the en-suite door or provision for it detailed to suit. See also *Hoists, design criteria* on page 15.

11.1.8 WC

A range of seat heights may be better provided by well selected fitting, such as one with seat thickness options, than ad-hoc plinths or seat adaptations. Soil outlet should make some provision for horizontal adjustment of WC to suit hoist track. Selection of WC should allow for shower chair use over bowl. The flushing lever should be large and sited on outer or transfer side. Capped electric supply for future Closomat or similar should be provided.

Figure 11.1.8

11.1.9 Shower

Whether a mixer fitting is installed at the outset or capped services provided, the shower should ensure ease of use from within the shower area. The drained area should be able to contain a shower chair and not constrict movement, say 1000 × 1000 mm as minimum.

Containment of water is often important, since a dry area outside the shower area is very desirable. Well sealing, half- or full-height easily operated doors may be advisable but should not restrict access or internal movement. Falls to floor outlets should be as slight as practicable and two-way falls should preferably be avoided for greater safety and to facilitate transfer. A level grating over a recessed tray may be a good option. Ramps must be avoided.

Controls should be safe, prevent scalding and be easily understood and manipulated. Control of water temperature and volume of flow should preferably be with one hand, whether push button or lever. There should be provision for a range of shower-head positions and they should be easily managed. Capped-off services should include supply for electric shower.

Provision for shower fitting to operate when bath is installed would be desirable (see 11.1.2).

Figure 11.1.9

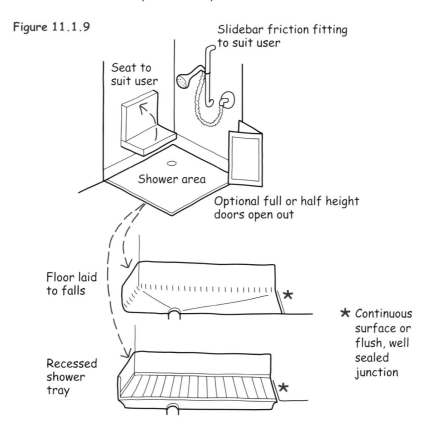

Slidebar friction fitting to suit user

Seat to suit user

Shower area

Optional full or half height doors open out

Floor laid to falls

Recessed shower tray

∗ Continuous surface or flush, well sealed junction

Profile of a wheelchair user 15

An older man living in a two-bedroom, ground floor, adapted, local authority flat with his wife, who has multiple sclerosis. The flat shares a common entrance hall.

The man has a manual wheelchair, and he and his wife both use scooters which are kept in the outside yard. He is unable to use his wheelchair indoors unless he removes the footrests. The shower-only bathroom has a sliding door.

There are parking problems. The flat is not very near to the parking area and there is no guaranteed parking space beside the kerb. Shopping is also a problem as the shops are too far away for reasonable access.

The front door has an intercom. The kitchen has had minor adaptations. The couple share tasks using a low-hinged table. The man dislikes having to ask neighbours to help him to change light bulbs.

11.1.10 Bath

Taps should be mounted on the outer side, provided that they do not hinder transfer, and must be easily manipulated, eg lever type. Recessed taps may be specified as an adaptation to suit an individual. Similarly, fitted side grips may be useful. The standard bath specified should, by design and strength, accept bath seats and bath boards. Bath lifts are not really practicable in shorter baths, 1700 mm is the preferred length. Good full length toe space should be provided. It should be possible to form a larger space between bath legs to accept a mobile hoist. Capped-off electrical supply for bath hoists should be provided.

Figure 11.1.10

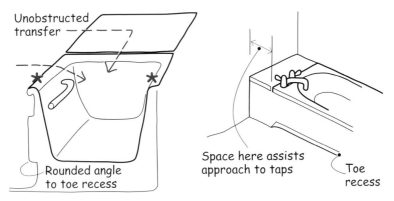

Unobstructed transfer

Rounded angle to toe recess

Space here assists approach to taps

Toe recess

★ Rims adequate to take bath board and provide hand support

11.1.11 Basin

The basin should allow a good approach, have a reasonable but shallow bowl area and have space around it for toiletries. Some adjustability within the fitting or by the vanity top is desirable. A specifically adjustable fitting may be required to suit an identified growing disabled child, but this would be an adaptation. Flexible

Profile of a wheelchair user 16

A woman with teenage children. She lives in her own adapted single-storey house, which has no transfer or re-charging space. She has an electric wheelchair. She finds the entry phone very helpful, but cannot close the door behind her. A remote door opener/closer would help her considerably. At present, she has to go through a series of complicated manoeuvres which are aggravated by the lack of a turning space inside the entrance.

The woman can pass through the bedroom door (measuring 925 mm wide) without difficulty, and can turn in the 1000-mm-wide passage with the wheelchair's footrests on. Other doors in the house are narrower and only passable with footrests off.

The kitchen has been adapted slightly. The worktop with sink has been lowered to 800 mm, with space available underneath.

plumbing should be considered. Fully inset or semi-recessed basins may restrict access, and full knee space underneath may be better. Taps may be separate or a mixer fitting, in which case a swivel arm would be helpful. Taps should be easily manipulated, eg lever type. As with kitchen sinks, provision for limited reach should be considered.

Figure 11.1.11

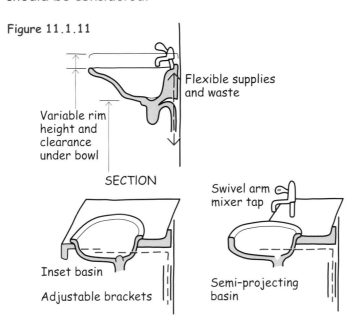

Flexible supplies and waste

Variable rim height and clearance under bowl

SECTION

Swivel arm mixer tap

Inset basin

Adjustable brackets

Semi-projecting basin

11.1.12 Supports

Fittings, including wall fixings, should be researched and specified for robustness. They may well act as supports in use and transfer, in particular basins and WCs.

Walls and ceilings should be able to accept a range of supports – rails, shower seat, full height poles, ceiling pulleys – all installed to individual requirements. See also *Hoists, design criteria* on page 15.

11.1.13 Heating

Radiators or heated towel rails should be carefully positioned so as not to compromise effective space and not to be available for support. Radiators should be of low surface temperature or protected.

11.1.14 Floor finishes

The floor finish in bathrooms should be slip resistant. Care should be taken over specification as some slip-resistant finishes can be very difficult to keep clean.

Wheelchair users may take weight on their feet, however momentarily, when using a shower for instance or leaving a bath. Slip resistance may also ensure the stability of wheeled equipment.

11.1.15 Wall finishes

Wall tiling should take account of extra splashing around fittings. All adjustable fittings should avoid disturbance to tiling.

Figure 11.1.15

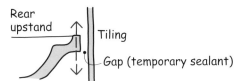

11.1.16 Fitting out

Provision should be made for toilet roll holders, towel hooks, mirrors and soap dishes, and cabinets should be fitted to suit individual reach and requirements.

11.1.17 Adaptations

Design and detail of space, services and structure should enable other possible adaptations, for example:

- specialist bath with swivel transfer seat
- shower bed over bath or shower area
- specialist WC with washing and drying functions, eg Closomat

Figure 11.1.17

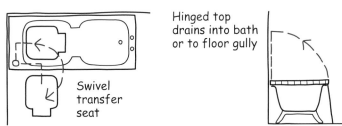

11.2 Requirements

11.2.1 Bathroom

In all dwellings provide fully accessible bathroom with WC, basin, and installed level-access shower with provision for bath in place of shower, with flexible or easily adapted services.

11.2.2 Access from bedroom

Ensure provision for direct access from main bedroom into bathroom.

11.2.3 Second WC

In dwellings of four or more persons, provide fully accessible second WC with basin, and hand the transfer space opposite to the handing of the main WC to provide both left-handed and right-handed transfer options within the dwelling.

Figure 11.2.3

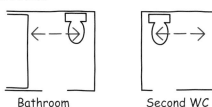

Bathroom   Second WC

11.2.4 Layout

Ensure independent approach/transfer to and use of all fittings, including manoeuvring space clear of fittings and of door swing if inward opening.

11.2.5 WC

Select and position for range of transfers and for provision of support rails later to suit user.

Figure 11.2.4/11.2.5

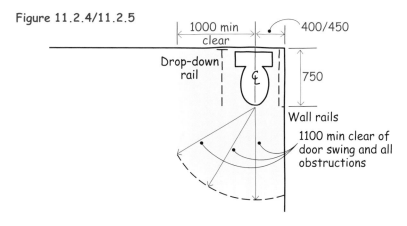

11.2.6 Shower

Where fully installed, detail to be fully accessible comprising drained floor, reachable and usable controls and scope for suitable water containment or suitable enclosure.

Figure 11.2.6

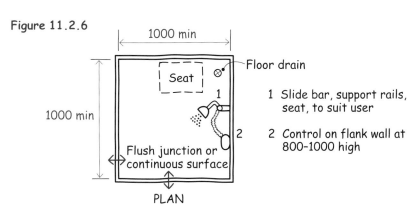

11.2.7 Bath

Where provided select bath and taps, position and detail to allow a range of transfers, access to and operation of taps.

Figure 11.2.7

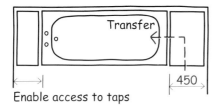

Enable access to taps

11.2.8 Basin

Select and position to be approachable in wheelchair with shallow but good capacity bowl, support for arms, reachable and usable taps.

Figure 11.2.8

750–800 rim height

600 clearance under bowl

11.2.9 Finishes

Provide adequate wall tiling and slip resistant flooring.

11.2.10 Supports

Ensure that walls and ceiling are adequate for adjustable height basins and subsequent fixings of hoists, seats, supports and other fittings.

Profile of a wheelchair user 17

A severely physically disabled young boy lives with his parents, sister and brother. They live in a bungalow adapted to provide covered transfer and access. He has a bedroom/activity room and en-suite bathroom. He is likely to use an electric wheelchair in the future.

In the bathroom, he transfers from the bed to the automatic wash/dry toilet and bath by overhead track and connecting doors. The WC outlet can be adjusted to allow final positioning to suit the track. There is a hinged showerbed over the bath. The basin is positioned in a vanity top on a Ropox frame, allowing fine adjustment between 650 mm and 910 mm, with taps placed at the front.

11.3 Recommendations

11.3.1 Second WC

In all homes provide fully accessible second WC with basin.

11.3.2 Shower

In second WC increase side transfer space to incorporate defined wheel-in shower area with floor drain.

Figure 11.3.2

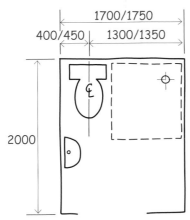

11.3.3 Fittings

Make spatial and/or serviced provision for subsequent adaptations to suit specific users.

WC:

- adjustable position to suit hoist track

- range of seat heights

- fitting with washing and drying functions

Bath:

- shower bed over bath

- specialist bath with integral transfer

Basin:

- height adjustable basin or counter top

12 Using bedrooms

Principle

Ensure that there is space in all bedrooms to accommodate the normal range of bedroom furniture, and for the wheelchair-using member of the household to enter, approach and transfer to beds, approach and use other furniture and operate windows.

12.1 Design considerations

12.1.1 Direct access from the main bedroom into the bathroom should be made possible whether a door is incorporated from the outset or its future provision is allowed for by means of a knock-out panel.

12.1.2 Basic provision should be a double bedroom and a single one. Single people often need a double bed. A second bedroom is necessary for family visitors, carers and extra equipment.

12.1.3 An effective double bedroom layout should allow a wheelchair user to:

- enter, manoeuvre clear of door swing, approach all furniture, leave room

- approach both sides of a double bed at an angle to transfer or attend to a child without need to reverse around the end of the bed

- access all electrical controls

- approach and control window

A view out of the window from the bed is desirable.

Figure 12.1.3

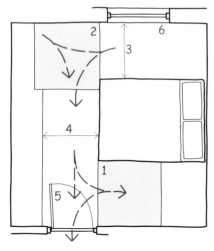

1 1200 x 1200 min activity square clear of bed, door swing and other fittings
2 1200 x 1200 min activity square clear of bed and other fittings
3 1000 min to approach bed and transfer
4 1000 min, governed by need to access wardrobes and drawers
5 Door to open beyond 90°
6 Window positioned for access to operate controls and to allow views out from bed position

12.1.4 The layout of single bedrooms should follow a similar pattern but access to one side of the bed is acceptable provided alternative bed and furniture layouts are possible which allow handed transfer to the opposite side of the bed.

12.1.5 Locating the door away from the corner of the room may assist manoeuvre. Facility to open the door beyond 90° may be helpful with manoeuvring, particularly on leaving.

12.1.6 Cupboards should be conveniently approached sideways or at a slight angle. Built-in cupboards may help to achieve a good layout.

12.1.7 Provision for a subsequent hoist installation should be made in all rooms for maximum flexibility of occupation. This may necessitate strengthening or detailing the bedroom ceiling to suit a hoist track. A track may be fitted to facilitate transfer between bed and chair or it may run from bedroom to bathroom. In the latter case it may run through the intervening wall or across circulation space and through doorways on line; track switching may be necessary. Provision for a connecting door can take several forms from knock-out panel to full provision from the outset, but door will need to be full height if hoist is installed. See also *Hoists, design criteria* on page 15.

12.1.8 Switches, sockets for equipment and for TV and FM aerials, a telephone and entry phone point should all be adjacent to the bed. This will determine optimum position of the latter within the room. A pull switch to operate the room light is desirable for ease of reach from the bed. See also Section 15 *Controlling services*.

12.2 Requirements

12.2.1 Layouts

Provide bedroom layouts to ensure access to both sides of beds in double bedrooms and outer side of beds in single bedrooms, access to other furniture and to window.

12.2.2 Controls

Adjacent to bedhead to provide:

- TV and FM aerial and power socket outlets

- room light switch, two-way with door switch

- entry phone point

- telephone point

Figure 12.2.2

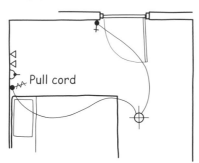

12.2.3 Door

Make provision for connection between main bedroom and bathroom by means of full-height knock-out panel, a door with panel over in full-height frame or fully detailed door.

12.2.4 Hoist

Make provision for future hoist installation in main bedroom – strengthen ceiling to allow run into bathroom, provide conduit and provide electrical wiring in roof space over.

12.3 Recommendations

12.3.1 Second or spare bedroom

Provide in all dwellings, layout allowing transfer to bed from either right or left side.

12.3.2 Hoist

Make provision for future hoist installation in other bedrooms from bed to chair space only, as 12.2.4.

13 Operating internal doors

Principle

Ensure that all internal doors, including those to storage spaces, can be operated conveniently.

13.1 Design considerations

13.1.1 Door handles may be levers with mortice latches, or pulls with adjustable ball catches. Pull handles can be extended in width to assist in closing the door when passing through, which will avoid the need for an additional closing pull in future. Large lever handles should always be preferred to knobsets for good grip. See also Section 4 *Negotiating the entrance door.*

Figure 13.1.1

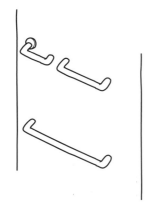

13.1.2 A fixing height for handles in the range 800–1000 mm is acceptable but 800 mm will assist those who need the support of the other arm or a wheelchair arm rest. This height would not inconvenience others. Alignment with adjacent controls such as switches is helpful as is contrast with the door itself.

Figure 13.1.2

13.1.3 The need for locking devices should be carefully considered. Where provided they should be easily manipulated.

13.1.4 It should be possible to release doors locked on the room side and to open doors to bathrooms and WCs outwards in emergency. Careful selection of fittings is important to allow easy operation by a wheelchair user if necessary.

Figure 13.1.4

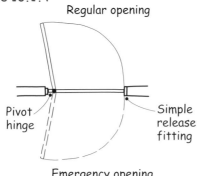

Regular opening

Pivot hinge

Simple release fitting

Emergency opening

13.1.5 Internal doors which are required to be fire resistant and self closing should be detailed for independent access bearing in mind that some wheelchair users may find operating a door difficult. Such doors can be held open or be swing free but at high cost. The best course is to select closers for minimum opening pressure and closing force, ie 15 newtons, and good range of adjustment of closing time and speed. The provision for installation of a remote-controlled door opener or assisted opening device as a future adaptation should be considered.

13.1.6 Well rounded exposed angles to door frames will help to minimise damage and the consequent need for repair or the fitting of protective angles.

Figure 13.1.6

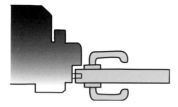

13.1.7 Sliding doors can have advantages:

- they do not intrude into room or circulation space when operated or left open

- they may be more easily operated by some wheelchair users

However, there are disadvantages which should be considered carefully:

- the need for wider structural opening to ensure clear opening width

- the need for space to slide door open, ie safeguarded wall space

- potential maintenance problems

- incompatibility with hoist tracks unless paired

Figure 13.1.7

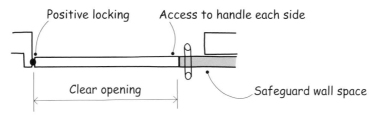

13.1.8 Where installed, sliding doors should have pull handles for convenient single-handed operation from either side, combined with locking where needed.

Figure 13.1.8

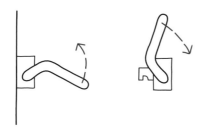

13.1.9 Reduced-swing doors, whilst not preferred, offer the limited advantages of convenient operation in either direction and saving on manoeuvring space to set against their higher cost and awkward operation. Fire-resisting self-closing versions are available, as are double-swing ones.

Figure 13.1.9

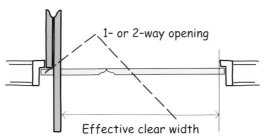

13.1.10 Double-swing doors, whether single to rooms or paired to suit a hoist track, should be considered carefully. Self-closing fittings such as spring hinges should provide gentle closing pressure, and door protection may be needed against damage from wheelchairs.

Figure 13.1.10

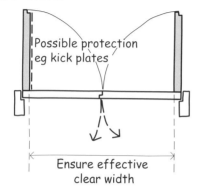

Possible protection eg kick plates

Ensure effective clear width

13.2 Requirements

13.2.1 Construction

Ensure that door construction permits subsequent fixing of pulls or other fittings.

13.2.2 Handles

Provide to all door types easily operated handles, pulls, latches and catches at 800–1000 mm high.

Figure 13.2.2

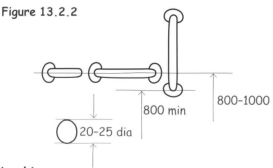

800 min

800–1000

20–25 dia

13.2.3 Locking

Ensure that locking/indicator devices are easily manipulated inside and outside in emergency.

13.2.4 Emergency opening

Ensure that inward-opening doors to bathrooms, WC and showers can be easily opened outwards in an emergency and by a wheelchair user.

13.2.5 Self-closing doors

Ensure, where these are required, independent operation from a wheelchair, with opening pressure not to exceed 15 newtons at the leading edge of the door.

13.3 Recommendation

13.3.1 Self-closing doors

Install or make provision for subsequent installation, as an adaptation, of door openers or assisted opening device operated by suitable remote control.

Profile of a wheelchair user 18

A young man living alone in a rented, one-bedroom, ground floor, adapted, local authority flat. He shares an entrance hall with other residents. He has three manual wheelchairs with interchangeable wheels and drives his own car. He needs to park away from the entrance and dislikes having to make the journey to and from his car in bad weather, especially if he is carrying shopping. He would like an integral garage or carport.

The door to his living room is badly sited making it difficult for the man to arrange furniture conveniently. The alternative external door is useful, but he is concerned that it is less secure than the main front door.

The bathroom has a shallow, low bath. The man would like a fitted shower for speed. He finds the bedroom too restricted. He would prefer a double bed, positioned against the flank wall and the window to maximise the available floor space.

For refuse disposal, he requires assistance to deposit waste disposal bags in the communal bin.

14 Operating windows

Principle

Ensure independent control of opening windows, passive and mechanical ventilation to requirements of Building Regulations and to reasonable level of comfort. Ensure balance of daylight, views out, privacy and security.

14.1 Design considerations

14.1.1 Selection, positioning and detailing of windows should satisfy a range of criteria as described below.

14.1.2 It is important to be able to see out from a seated position; horizontal framing at this lower eye level should be avoided. Too-high sills may restrict views of ground and plants immediately outside the windows.

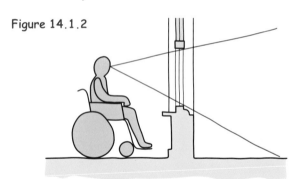

Figure 14.1.2

14.1.3 Windows to floor level or low level are beneficial for views out, but if excessive will seriously restrict furniture and radiator layouts. They may require excessive curtaining and may engender feelings of insecurity or over-exposure in certain locations, eg facing a street or footpath.

Profile of a wheelchair user 19

A young woman, who was disabled as a result of a car accident, lives with her husband in their own adapted and extended bungalow. She uses a manual wheelchair and drives a car. The ability to transfer under cover is important to her. She uses the adjoining garage.

She exercises in the living room using a standing frame which is usually stored outside the room. She works on her home computer, hopes to have children, and is extending her home to provide accommodation for two children and a live-in carer. She is planning to replace the present WC/shower with a bathroom containing a specialist bath with transfer seat, and a separate wheel-in shower area.

14.1.4 In addition the safety and protection of glazing is covered by Building Regulations requirements and provisions. The need for and cost of safety glazing below 800 mm should be set against the need for good views out. In practice a glazing height of 800 mm should satisfy most users' need to see out comfortably.

Figure 14.1.4

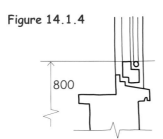

14.1.5 For those with limited mobility, large glazed areas may also cause discomfort from glare. Take care over orientation especially to south-west and west. Less mobile occupants might also value wider sills on which to place things, provided that these do not impede approach to controls.

Figure 14.1.5

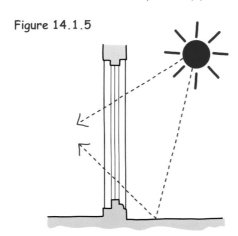

14.1.6 Bay windows or windows on adjoining walls increase the range of views out for those with limited mobility and may help to bring daylight into deep plans.

Figure 14.1.6

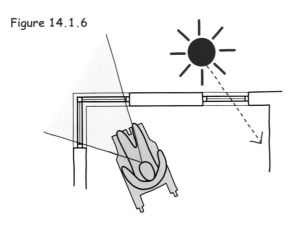

14.1.7 The scope for cleaning windows should be carefully considered. Fully reversible windows can be difficult to operate but side swing or projecting casements of reasonable width, say 600 mm, offer the possibility of cleaning from inside.

14.1.8 Window fittings should be accessible from a wheelchair, possibly by a sideways approach. Aim to avoid the need for remote-control gear but provision for it as an adaptation may be needed. Kitchen and bathroom layouts should where possible avoid windows over worktops or deep fittings, and positioning in living areas and bedrooms should take account of likely furniture arrangements as far as possible. See also Section 12 *Using bedrooms*.

Figure 14.1.8

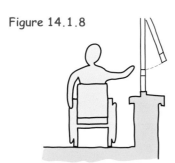

14.1.9 Window fittings should preferably be easily manipulated with one hand, providing controlled ventilation or wider opening and security in ventilation or closed positions. Top-hung or side-hung windows with conventional or reflex/projecting hinges may be the most suitable in this respect.

Figure 14.1.9

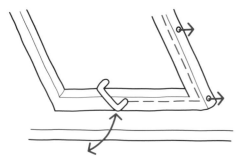

14.1.10 The need for manual or electrical opening should be borne in mind both in initial selection of appropriate windows and in provision as an adaptation for electrical control.

14.1.11 Where passive ventilation devices such as trickle vents are required they should be provided or duplicated at lower level for control from a wheelchair.

14.1.12 Mechanical ventilation should preferably be limited to satisfying Building Regulations. It should not be used to substitute for inaccessible windows where they could be avoided by careful layout.

14.1.13 Outward-opening windows over paths and gardens may cause a hazard particularly to wheelchair users or children.

Figure 14.1.13

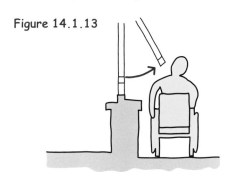

14.2 Requirements

14.2.1 Approach

Ensure that wheelchair user can approach window to operate controls for opening and ventilation.

14.2.2 Operate

Ensure single operating handle within reach, controlling both ventilation and full opening and closing positions while maintaining security in ventilation as well as closed positions.

Figure 14.2.2

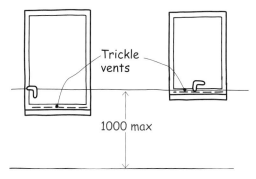

Trickle vents

1000 max

14.2.3 Gear

Where conventional opening is not possible, install manual or powered window-opening gear with accessible handle to control main or most appropriate opening.

Figure 14.2.3

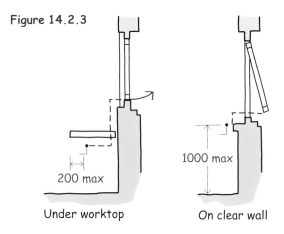

200 max

1000 max

Under worktop   On clear wall

14.2.4 Safety

Ensure that windows opening out over paths do not create hazards.

14.2.5 Glazing

Ensure that glazing line to windows in living, dining and bedrooms is no higher than 800 mm.

14.2.6 Transoms

Avoid full-width transoms (horizontal divisions) between 800 and 1500 mm high.

Figure 14.2.6

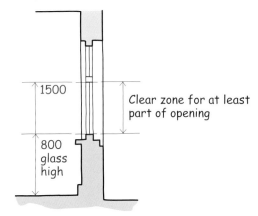

1500

Clear zone for at least part of opening

800 glass high

15 Controlling services

Principle

Ensure that all essential controls are accessible to a wheelchair user and easily manipulated or operated.

15.1 Design considerations

15.1.1 These should be used in conjunction with references to such controls within the preceding activities. Although it may come within the scope of specific adaptations, the need for controls which provide tactile or large-format information for visually impaired people should be considered. Consistency in heights and alignment with door handles should be aimed at with such users in mind.

15.1.2 Access to services may be facilitated by careful positioning and selection. It may in some cases be more easily facilitated by simple remote-control devices, for example of stop taps.

15.1.3 The need for ease of manipulation embraces all fittings including central heating controls. Light switches should be full plate type for ease of use by all, and double sockets should have switches to outer ends. Pull-cord switches should have larger pulls and restraining eyes to ensure that cord stays within reach.

Figure 15.1.3

 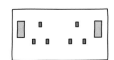

Full plate   Socket outlet   Pull cord
switch   switched at outer ends

15.1.4 Height of sockets and switches and other wall-mounted controls should be set by acknowledging severely limited reach or arm strength of some users. Controls should be within the range 700–1050 mm. It may be desirable to align sockets and switches in height, at say 900 mm. See also Section 10 *Using the kitchen* for flexible controls.

Figure 15.1.4

15.1.5 Switches should always be conveniently sited and two-way switching should be installed, for example, to external routes (carport and entrance, external store and secondary entrance) and internally (internal and external doors in living rooms, bed and door in bedrooms).

15.1.6 Security from intruders concerns many wheelchair users and a specific and effective intruder alarm installation is a desirable initial installation.

15.1.7 Being able to replace light bulbs or lamps concerns some wheelchair users on grounds of safety and personal independence. These concerns should be addressed as far as practicable. Possible responses could include the following:

- limit need for artificial lighting by good standard of distributed daylight especially in critical areas such as entrance halls and passages, bathrooms and kitchens

- select light fittings with long-life lamps

- provide emergency or fail-safe lighting in critical areas

- make good provision for sockets in living rooms for wall or standard lights

- provide local kitchen lighting within reach

15.1.8 Room space heating should be responsive and controllable and achieve even temperatures throughout the house and at required times, with accessible thermostatic valves to provide up to 22 °C local control. Some wheelchair users need above-average temperatures for personal comfort and may be susceptible to a drop in temperature as they move through their home.

15.1.9 Low-surface-temperature radiators or heaters should be considered, particularly in confined areas such as passages, bathrooms and WCs.

15.1.10 Hot water temperatures should not exceed 43° at any fitting.

15.1.11 Flexible pipework should be provided to all adjustable fittings such as sink and basin.

15.1.12 Telephone installation should serve entrance, kitchen, living area and all bedrooms and can incorporate emergency and alarm facilities.

15.1.13 Electrical installations should reflect the high level of current domestic demand for control of services, appliances, communication and entertainment systems, alarm and environmental systems. They should also take into account the increased needs of wheelchair users, particularly those with limited reach and mobility, and include provision for the adaptations, such as door operators, identified in the earlier activities.

Where increased provision of sockets may be needed – for computers or entertainment systems – consider recessed wall-mounted trunking with scope for added sockets. Highly personalised environmental controls are, however, a matter for the one-off dwelling or specific adaptation for the individual wheelchair user.

15.2 Requirements

15.2.1 Main services

Ensure that a wheelchair user can reach, control and read the following:

- mains water stopcock
- gas and electricity main switches and consumer units

15.2.2 Plumbing

Provide essential isolating stop taps to sink, washing machine, WC and shower, and ensure that control by wheelchair user is possible.

Figure 15.2.2

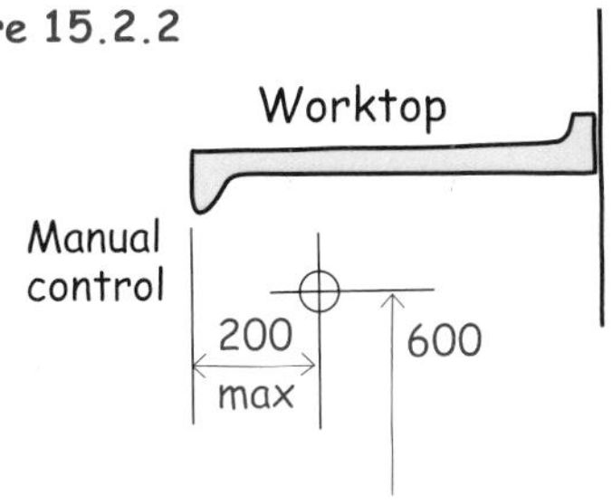

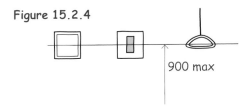

15.2.3 Flexible plumbing

Provide flexible plumbing to all adjustable fittings.

15.2.4 Switches

Specify full plate or large rocker light switches, two-way where required, and set at 900 mm high maximum.

Where pull switches are required, provide large pull at 900 mm high maximum and cord-restraining eye.

Figure 15.2.4

15.2.5 Socket outlets: general

Specify sockets with large switches, on outer ends of double sockets.

Set sockets generally at 700 mm high minimum. Where they are above worktops set sockets at 100 mm above worktop level. Where worktop height is adjustable, fit laboratory-type sockets to worktop.

Figure 15.2.5

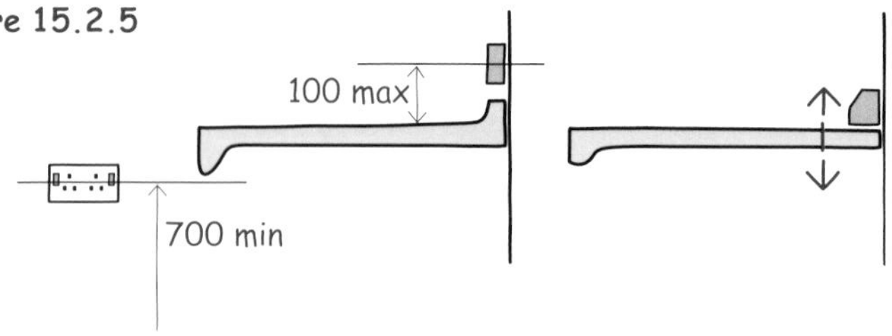

100 max

700 min

15.2.6 Socket outlets: appliances

Set remote switches generally at 900 mm maximum. Where above fixed worktops, set switches 100 mm above worktop level. Where adjustable, fit to worktop as 15.2.5.

Set sockets served by remote switches at 600 mm minimum where they are below worktops.

Figure 15.2.6

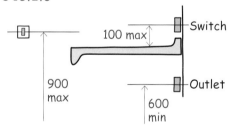

100 max — Switch

900 max

Outlet

600 min

15.2.7 Telephone

Provide conduit or draw cord or line in to outlet sockets in entrance, living room, kitchen, main bedroom and upper level circulation, all at 700 mm or to match general socket outlets.

15.2.8 Future controls

Provide supply to future powered elements identified as *Requirements* in previous activities, such as entry phone and door openers.

15.2.9 Heating

Ensure that heating controls, such as boiler ignition, programmer/timer/pump, thermostat, are within reach and easily read and operated.

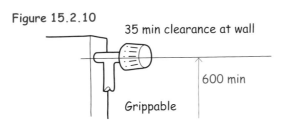

Figure 15.2.9

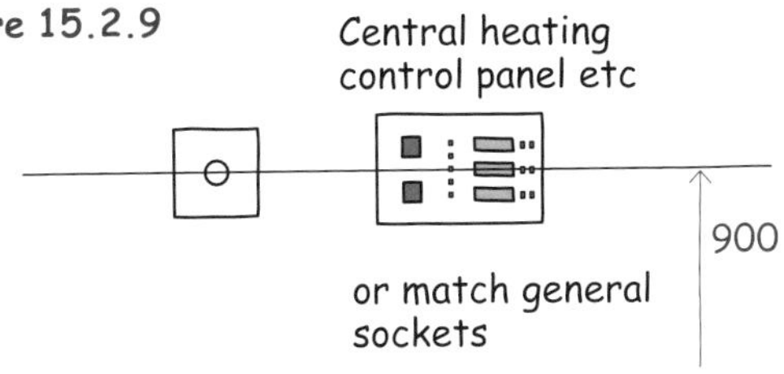

Central heating control panel etc

or match general sockets

900

15.2.10 Radiators, heaters

Site radiator valves and heater controls at 600 mm high minimum, ie mounted at top of radiator, and specify usable fittings.

Figure 15.2.10

35 min clearance at wall

600 min

Grippable

15.2.11 Surface temperatures

Install low-surface-temperature radiators in bathrooms, WCs and restricted circulation areas. Ensure that thermostatic valves are outside casing.

15.2.12 Water temperatures

Ensure that hot water temperature does not exceed 43° at any fitting.

15.2.13 Security

Make suitable provision for personal alarm, fire alarms and for installation of intruder alarm by occupant.

15.3 Recommendations

15.3.1 Switches and sockets

Align both at 900 mm height, other than sockets over or under worktops.

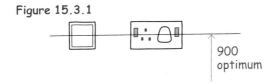

Figure 15.3.1

900 optimum

15.3.2 Radiator controls

Set at 800 mm optimum height.

Figure 15.3.2

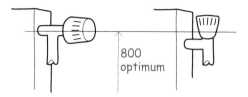

15.3.3 Socket switching

Where ability to reach wall-mounted switched sockets could be problematic, such as above kitchen worktops, install fascia-mounted switches or make provision for an adaptation appropriate to the user.

Figure 15.3.3

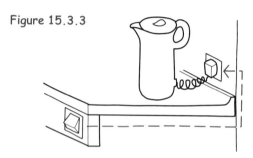

15.3.4 Lighting

Make optimum practical provision for replacement of lamps by wheelchair user and minimise need or ensure safety by long-life lamps or emergency lighting.

15.3.5 Radiators

Install low-surface-temperature radiators throughout.

15.3.6 Future controls

Provide supply to future powered elements identified as *Recommendations* in previous activities.

Appendix A: Summary of *Requirements*

Have you satisfied the following *Requirements* where they apply to your scheme?

This is purely to enable you to check that your scheme complies. You should have referred to the appropriate item for detailed, illustrated guidance when designing and you may need to refer again to check.

Moving around outside
1.2.1	**Footpaths:** space and safety?
1.2.2	**Footpath edges:** protection?
1.2.3	**Gradients:** lengths, landings, handrails?
1.2.4	**Cross falls:** minimum falls?
1.2.5	**Crossings:** barrier free?

Using outdoor spaces
2.2.1	**Gardens:** gates?
2.2.2	**Balconies:** accessible and usable?
2.2.3	**Garden:** access?
2.2.4	**Refuse:** manageable?

Approaching the home
3.2.1	**Direct entrance parking:** adjacent?
3.2.2	**Covered area:** area and height?
3.2.3	**Communal entrance parking:** designated?
3.2.4	**Garage:** area and height?
3.2.5	**Route to entrance:** surface, gradients?
3.2.6	**Entrance landing:** level area, protection?
3.2.7	**Canopy at entrance:** extent and height?
3.2.8	**Lighting:** adequate, detection?
3.2.9	**Lifts:** provision, capacity?

Negotiating the entrance door
4.2.1	**Door:** effective clear width?
4.2.2	**Approach space:** provision?
4.2.3	**Threshold:** functional?
4.2.4	**Lock:** secure?
4.2.5	**Opener:** provision?
4.2.6	**Lock, common external door:** opener?
4.2.7	**Lever, pull handles:** grip, contrast?
4.2.8	**Entry phone:** siting?
4.2.9	**Bell:** siting?
4.2.10	**External light:** effective, detection?
4.2.11	**Lighting:** controls?
4.2.12	**Pull:** provision?

Entering and leaving; dealing with callers

5.2.1   **Transfer:** adequate space?
5.2.2   **Turning space:** adequate space?
5.2.3   **Post:** fitting?
5.2.4   **Entry phone:** provision?
5.2.5   **Lobby:** adequate space?

Negotiating the secondary door

6.2.1   **Landing:** level, adequate space?
6.2.2   **Door:** effective clear width?
6.2.3   **Approach space:** adequate space?
6.2.4   **Threshold:** functional?
6.2.5   **Lock:** security and operation?
6.2.6   **Stays:** provision?
6.2.7   **Pull:** provision?
6.2.8   **Lighting:** provision, detection?

Moving around inside; storing things

7.2.1   **Passages:** adequate width?
7.2.2   **Doors in passages:** adequate space to approach?
7.2.3   **Turning 90°:** adequate space?
7.2.4   **Turning 180°:** adequate space?
7.2.5   **Right angles:** adequate space?
7.2.6   **Effective clear door widths:** opening width?
7.2.7   **Space to approach doors:** adequate space?
7.2.8   **Doors at angles:** adequate space to turn?
7.2.9   **Sliding doors:** adequate space?
7.2.10   **Storage:** usable?

Moving between levels within the dwelling

8.2.1   **Lift:** provided?
8.2.2   **Installation:** secure and safe?
8.2.3   **Circulation:** adequate space?

Using living spaces

9.2.1   **Room layout:** adequate usable space, furniture?
9.2.2   **Radiators:** positioning?
9.2.3   **Sockets:** positioning?

Using the kitchen

10.2.1   **Layout:** practical for a wheelchair user?
10.2.2   **Worktops:** adequate provision?
10.2.3   **Sink:** suitable?
10.2.4   **Storage:** usable provision?
10.2.5   **Controls:** suitable provision?
10.2.6   **Appliances:** adequate provision?
10.2.7   **Refuse:** usable provision?

Using the bathroom

11.2.1   **Bathroom:** fully accessible, bath, shower provision?
11.2.2   **Access from bedroom:** provision?
11.2.3   **Second WC:** provided, handed?
11.2.4   **Layout:** independent use?
11.2.5   **WC:** appropriate?
11.2.6   **Shower:** details, usable?
11.2.7   **Bath:** details, usable, transfers?
11.2.8   **Basin:** approachable, usable?
11.2.9   **Finishes:** adequate provision?
11.2.10 **Supports:** adequate provision?

Using bedrooms

12.2.1   **Layouts:** accessible?
12.2.2   **Controls:** provision and positioning?
12.2.3   **Door:** en-suite provision?
12.2.4   **Hoist:** provision?

Operating internal doors

13.2.1   **Construction:** provision for fittings?
13.2.2   **Handles:** operation?
13.2.3   **Locking:** operation?
13.2.4   **Emergency opening:** provision, operation?
13.2.5   **Self-closing doors:** operation, pressure?

Operating windows

14.2.1   **Approach:** approachable?
14.2.2   **Operate:** control of ventilation security?
14.2.3   **Gear:** provision where necessary?
14.2.4   **Safety:** window positioning?
14.2.5   **Glazing:** height?
14.2.6   **Transoms:** positioning?

Controlling services

15.2.1   **Main services:** controls, reading?
15.2.2   **Plumbing:** controls at fittings?
15.2.3   **Flexible plumbing:** provision?
15.2.4   **Switches:** specification and positioning?
15.2.5   **Socket outlets: general:** specification and positioning?
15.2.6   **Socket outlets: appliances:** positioning of switches, sockets?
15.2.7   **Telephone:** positioning?
15.2.8   **Future controls:** provision?
15.2.9   **Heating:** controls?
15.2.10 **Radiators, heaters:** controls?
15.2.11 **Surface temperatures:** vulnerability?
15.2.12 **Water temperatures:** controlled, maximum?
15.2.13 **Security:** provision?

Appendix B: Checklist of best practice

Reference numbers are to *Design considerations*.

This is purely to enable you to check that your design incorporates the optimum level of usability. You should have referred to the appropriate item for detailed, illustrated guidance and you may need to refer again to check.

Moving around outside

1.1.2	Have you checked out ease of movement off site?
1.1.3	Where are the basic facilities?
1.1.4	Have you ensured ease of movement within the site?
1.1.5	Are the routes accessible, safe and secure?
1.1.6	Have you taken care over slopes and gradients?
1.1.7	Have you taken care over cross falls?
1.1.8	Are the paved surfaces suitable?
1.1.9	Is there defensible space in front of dwellings?
1.1.10	Have you investigated alternatives to single-storey dwellings?
1.1.11	Have you minimised distinctiveness of the dwellings?
1.1.12	Are all the dwellings visitable?

Using outdoor spaces

2.1.1	How is boundary between public and private spaces achieved?
2.1.2	Are the rear gardens safe, secure but accessible?
2.1.3	Are any balconies usable?
2.1.4	Are gardens laid out for access and enjoyment?
2.1.5	Is paving carefully detailed?
2.1.6	Have you ensured usability of refuse arrangements?
2.1.7	Are there usable clothes-drying facilities?

Approaching the home

3.1.1	Have you ensured effective cover where transferring between vehicles and negotiating the entrance door?
3.1.2	Is the carport carefully sited in relation to windows?
3.1.3	Is side protection to the carport needed and provided?
3.1.4	Have you allowed for a range of transfer methods?
3.1.5	What provision would there be for class 3 vehicles, ie scooters?
3.1.6	Have you resolved grouped parking issues?
3.1.7	Does the route to, and the space at, the entrance suit a wheelchair user?
3.1.8	Have you minimised the need for slopes and gradients?
3.1.9	Is lighting adequate and have you considered control by detectors?
3.1.10	Are the garages adequate for transfer inside them?
3.1.11	Have you resolved lift provision in multi-storey developments?

Negotiating the entrance door

4.1.2	Is the door wide enough?
4.1.3	Is there appropriate space to approach and operate the door where it opens towards the user?
4.1.4	Is there appropriate space where the door opens away from the user?
4.1.5	Do the thresholds satisfy all the functional requirements?
4.1.6	Can door locks be reached and operated?
4.1.7	Is the optimum level of security provided?
4.1.8	Are all fittings easy to grip, with fiddly operations avoided?
4.1.9	Are common entrance doors convenient to use?
4.1.10	Have entry phone systems been carefully selected?
4.1.11	Can a horizontal pull handle be added later?
4.1.12	Is the entrance with all its elements properly lit with good signing?
4.1.13	Are the light controls convenient with two-way switching where necessary?

Entering and leaving; dealing with callers

5.1.1	Have you satisfactorily resolved transfer between chairs, storage and charging, without compromising independence?
5.1.2	Can a user approach the door and manoeuvre to deal with post, callers and visitors?
5.1.3	Is there provision for a wheelchair-using visitor?
5.1.4	Is the post collection fitting positioned properly?
5.1.5	Have entry phone arrangements been made?
5.1.6	How is it possible to see who is calling?
5.1.7	Have you provided natural lighting?
5.1.8	Have you allowed for matting to be added?
5.1.9	Have you considered lobbies and if so do they work?

Negotiating the secondary door

6.1.1	Is this door carefully sited in relation to circulation or living space?
6.1.2	Is there adequate space to approach this door?
6.1.4	Have you considered door options carefully for access and security?
6.1.5	Have you resolved whether doors open in or out?
6.1.6	Have you resolved security and ease of exit?
6.1.7	Is lighting conveniently controlled internally and externally when using the door?

Moving around inside; storing things

7.1.1	Will circulation be adequate within the dwelling?
7.1.2	Is there space for all critical wheelchair manoeuvres?
7.1.3	Can you provide space beyond the minimum?
7.1.4	Have you ensured that minimum passage widths will be clear of added projections?
7.1.5	Will right angles be negotiable?

7.1.6	Have you incorporated splays to minimise damage and to improve circulation?
7.1.7	Have you got away from corridors?
7.1.8	Have you related door opening widths to the way they are approached and negotiated?
7.1.9	Can doors open beyond 90°?
7.1.10	Do doors opening off passages have at least 775 mm effective clear width?
7.1.11	Have you increased this width to suit doors opening into a passage?
7.1.12	Have you considered door opening width in relation to leaving the bathroom?
7.1.13–14	Can doors be approached and operated conveniently head-on?
7.1.15	Is it convenient to turn and negotiate doors at the end of a passage?
7.1.16	Have you considered alternatives to hinged doors?
7.1.17	Can sliding doors be approached for ease of operation?
7.1.18	Have you avoided need for added protection?
7.1.19	Have you provided daylighting?
7.1.20	Do dimensions of storage spaces and the way they are laid out ensure effective use in a wheelchair?

Moving between levels within the dwelling

8.1.1	Does the lift connect circulation areas at each level?
8.1.3	Have you researched choice of lift in relation to user need and space required?
8.1.4	Have you incorporated space to approach the lift and operate the door?
8.1.5	Is there a full range of safety provision?

Using living spaces

9.1.1	Have you carefully considered the relationship of living, dining and kitchen spaces appropriate to the size of the dwelling?
9.1.2–3	Will occupants have reasonable scope to lay out rooms as they wish?
9.1.4	Have radiator and socket positions been considered carefully?

Using the kitchen

10.1.1	Will a range of wheelchair users be able to manage the kitchen?
10.1.2	Have you arrived at an effective layout to suit the dwelling size?
10.1.3	Have you resolved sequences, continuity and travel distances?
10.1.4	Is there enough provision for a wheelchair user to approach under the worktop, hob, sink?
10.1.5	Have you avoided the need for excessive manoeuvring or positioning?
10.1.6	Are built-in appliances properly integrated?
10.1.7	Have you resolved the question of adjustable worktops?
10.1.8	Do you know the intending user at design stage?

10.1.9	Is there scope to experiment in arriving at the best height?
10.1.10	Is a compromise height needed?
10.1.11	Will wall brackets be clear of wheelchair approach?
10.1.12	Is there adjustability of different sections of worktop?
10.1.13	Can worktop height be adjusted without disturbance to wall tiling?
10.1.14	Have you resolved internal angles in terms of approach and reach?
10.1.15	Have you incorporated pull-out worktops?
10.1.16	Has the sink been selected carefully?
10.1.17	Have the taps been selected carefully and is there, or will there be, a need for separate controls to suit limited reach?
10.1.18	Is any clear space under the sink out of view from living areas?
10.1.19	Will there be enough reachable storage?
10.1.20	Have you considered the full range of possible storage solutions?
10.1.21	Are handles robust and grippable?
10.1.22	Is there provision for both built-in and added appliances?
10.1.23	Is there scope for individual selection of appliances to suit specific needs or preferences?
10.1.24	Has access to controls and switches been fully resolved in terms of limited reach and to suit adjustable worktops?
10.1.25	Are out-of-reach floor areas avoided by, for example, plinths?
10.1.26	Have other utility activities been considered and allowed for?

Using the bathroom

11.1.1	Will provision be appropriate?
11.1.2	Have you resolved shower, bath and WC provision, now and future?
11.1.3	Have you resolved the question of handing the WC for maximum usability?
11.1.4	Have you ensured appropriate space?
11.1.5	Have you based your layout on the detailed guidance set out here?
11.1.6	Have you considered carefully type of door, its position and the way it functions? Have you considered shower provision and a basin in the second WC?
11.1.7	Have you made provision for a connecting door from the main bedroom with provision for hoist track?
11.1.8	Have you resolved need for a range of WC seat heights?
11.1.9	Is the shower detailed in all respects to be usable, safe and easily controlled, and is containment of water resolved?
11.1.10	Has bath been selected for maximum usability – transfer, supports, access to taps, provision for hoist?
11.1.11	Has basin been selected for approach and use and has the need for adjustable height been established?
11.1.12	Are fittings and their fixings robust?
11.1.13	Have radiators been positioned carefully and any need for low surface temperatures established?

11.1.14　Have floor finishes been selected for slip resistance and cleanability?

11.1.15　Is extent of wall tiling adequate and are adjustable fittings detailed to avoid disturbing tiling?

11.1.16　Is there provision for the full range of accessories?

11.1.17　Have you made reasonable provision for adaptations including specialist fittings?

Using bedrooms

12.1.1　Is there provision for a connecting door to the bathroom and will it accept a hoist track?

12.1.2　Is there a second bedroom?

12.1.3–4　Is each bedroom laid out for appropriate access to beds and other furniture and all controls?

12.1.5　Is door away from corner and does it open beyond 90°?

12.1.6　Are cupboards positioned or detailed to be usable?

12.1.7　What provision has been made for subsequent hoist installation – layout, door positions, strengthening?

12.1.8　Are switches, sockets related to bed position?

Operating internal doors

13.1.1　Have handles been selected for good grip?

13.1.2　Are handles at a suitable height?

13.1.3　Has need for locking been established and any provision carefully detailed?

13.1.4　Has need for emergency opening been established and any provision carefully detailed?

13.1.5　Have you resolved the operation of any fire doors to avoid barriers to independent access?

13.1.6　Have you detailed door frames to minimise damage in use or need for added protection?

13.1.7　Have overall advantages of sliding doors been established and is their installation suitably detailed?

13.1.8　Can sliding doors be operated conveniently?

13.1.9　Have reduced-swing doors been considered?

13.1.10　Have double-swing doors been installed and are the self-closing fittings suitable?

Operating windows

14.1.1　Do the windows as selected satisfy the relevant criteria set out?

14.1.2　Are views out unhindered?

14.1.3　Have any low sill heights been considered very carefully?

14.1.4　Do selected sill heights balance views out, safety and the budget?

14.1.5　Have you avoided discomfort arising from size or positioning and have you considered sill details carefully?

14.1.6　Have you been able to incorporate bay windows or windows on adjoining walls?

14.1.7　Have you resolved the question of window cleaning?

14.1.8 Can all window fittings be reached?

14.1.9 Can all windows be easily manipulated to control ventilation and security?

14.1.10 Have you taken account in selection of future need for remote manual or electric opening?

14.1.11 Have trickle vents been provided at low level?

14.1.12 Have you restricted need for mechanical ventilation?

14.1.13 Have you taken care over outward-opening windows in relation to activities immediately outside?

Controlling services

15.1.1 Have you taken account of the needs of visually impaired people?

15.1.2 Have you resolved access to services and exploited remote-control possibilities?

15.1.3 Have all fittings been selected for ease of manipulation?

15.1.4 Are all sockets and switches at a suitable height?

15.1.5 Have you provided two-way switching wherever it will help the user?

15.1.6 Have you carefully considered and resolved the range of safety and security needs?

15.1.7 Have you resolved concerns about failure and replacement of light bulbs?

15.1.8 Will room space heating be satisfactory, responsive and controllable?

15.1.9 Has the need for low-surface-temperature radiators been established?

15.1.10 Are maximum hot water temperatures controlled?

15.1.11 Do all adjustable fittings have flexible pipework?

15.1.12 Does telephone installation serve different parts of the dwelling and incorporate emergency facilities?

15.1.13 Does electrical installation take account of present high level of demand, increasing needs of wheelchair users and future provision?

'Secured by design'

Achievement of the standards set by the 'Secured by design' initiative should be possible without compromising the principles of independent access and management set out in this guide.

In some cases the guidance on access will reinforce security, for example that on external lighting and parking arrangements. The critical elements which may need an agreed resolution at design stage are the detailing of **entrance** and other external doors to the dwelling, and of **windows**, in both cases to achieve an optimum balance between security and usability.

Bibliography

British Standards

BS 6440:1999 Powered lifting platforms for use by disabled persons. Code of practice.

BS 7036-2:1996 Code of practice for safety at powered doors for pedestrian use. Part 2. Straight and curved sliding doors and prismatic and folding doors

BS 7036-3:1996 Code of practice for safety at powered doors for pedestrian use. Part 3. Swing doors and balanced doors

BS 7036-4:1996 Code of practice for safety at powered doors for pedestrian use. Part 4. Low energy swing doors

BS 8220-1:2000 Guide for security of buildings against crime. Part 1. Dwellings

BS 8300:2001 Design of buildings and their approaches to meet the needs of disabled people. Code of practice

College of Occupational Therapists, Housing Corporation. *Minor adaptations without delay: a practical guide and technical specifications for housing associations.* London, College of Occupational Therapists Ltd, 2006. Also available on the website *http://www.cot.org.uk/newpublic/about/minor_works.php*

Department of the Environment, Transport and the Regions. *Guidance on the use of tactile paving surfaces.* London, DTLR Mobility and Inclusion Unit, 1999.

Department of the Environment, Transport and the Regions. *Accessible thresholds in new buildings: guidance for house builders and designers.* London, The Stationery Office, 1999.

Garvin S L. *Domestic automatic doors and windows for use by elderly and disabled people: a guide for specifiers.* BRE Report BR334. Garston, BRE Press, 1997.

Habinteg Housing Association. *Design guide.* London, Habinteg, 2003.

Habinteg Housing Association. *Realities of independent living: supported housing design guidance.* London, Habinteg, 2005.

Joseph Rowntree Foundation. *Meeting Part M and designing lifetime homes.* York, JRF, 1999.

National Disabled Persons Housing Service Ltd (HoDis). *Needs first: a good practice guide for RSLs prioritising tenants needs for adaptations.* York, HoDis, 2001.

National Housing Federation. *Standards and quality in development: a good practice guide.*

Office of the Deputy Prime Minister. *Delivering housing adaptations for disabled people: a good practice guide.* Wetherby, ODPM Publications, 2004. Also available on the website *http://www.odpm.gov.uk/index.asp?id=1152864*

Office of the Deputy Prime Minister. *Planning and access for disabled people: a good practice guide.* Wetherby, ODPM Publications, 2003. Also available on the website *http://www.odpm.gov.uk/index.asp?id=1144644*

Office of the Deputy Prime Minister. *The Building Regulations 2000 Approved Document M. Access to and use of buildings* (2004 edition). London, The Stationery Office, 2004.

Parker P and Fraser J. *Sign design guide: a guide to inclusive signage.* JMU Access Partnership and The Design Society, 2000. ISBN 185878 412 3.

Royal National Institute for the Blind (RNIB). *A design guide for the use of colour and contrast to improve the built environment for visually impaired people.* RNIB/GDBA Joint Mobility Unit, University of Reading, Research Group for Inclusive Environments, Department of Construction Management and Engineering, Whiteknights, PO Box 219, Reading, RG6 6AW; and ICI plc. 1997.

Royal National Institute for the Blind (RNIB). *Building sight: a handbook of building and interior design solutions to include the needs of visually impaired people.* London, The Stationery Office, 1995.

Other sources of information

Assist UK

Assist UK leads a network of local Disabled Living Centres. Each centre includes an exhibition of products and equipment and offers information on access, design and equipment for daily living.

Redbank House, 4 St Chad's Street, Manchester M8 8QA
Tel: 0870 770 2866, Textphone: 0870 770 5813, Fax: 0870 770 2867
E-mail: general.info@assist-uk.org, www.assist-uk.org

Centre for Accessible Environments

CAE is concerned with the practicalities of inclusive design in the built environment and provides information, design guidance, training and consultancy services.

70 South Lambeth Road, London SW8 1RL
Tel/textphone: 020 7840 0125, Fax: 020 7840 5811, SMS 07921 700098
E-mail: info@cae.org.uk, www.cae.org.uk

College of Occupational Therapists

The professional, educational and trade union organisation for occupational therapists, their support staff and students in the UK.

106–114 Borough High Street, Southwark, London SE1 1LB
Tel: 020 7357 6480
www.cot.org.uk

Disability Rights Commission

An independent body established in 2000 by Act of Parliament to stop discrimination and promote equality of opportunity for disabled people.

Tel: 08457 622 633, Textphone: 08457 622 644, Fax: 08457 778 878
DRC Helpline, Freepost MID02164, Stratford upon Avon CV37 9BR

Habinteg Housing Association Ltd

Habinteg is the UK's leading expert in accessible housing and disability.

Holyer House, 20–21 Red Lion Court, London EC4A 3EB
Tel: 020 7822 8700, Fax: 020 7822 8701
Email: info@habinteg.org.uk, www.Habinteg.org.uk

Joseph Rowntree Foundation

This social policy research and development charity spends about £7 million a year on a research and development programme that seeks to better understand the causes of social difficulties and explore ways of overcoming them.

The Homestead, 40 Water End, York, North Yorkshire YO30 6WP
Tel: 01904 629241, Fax: 01904 620072
Email: info@jrf.org.uk, www.jrf.org.uk

National Housing Federation

The Federation represents 1400 independent, not-for-profit housing associations in England, who provide over 2 million homes for 4 million people.

Lion Court, 25 Procter Street, London WC1V 6NY
Tel: 020 7067 1010, Fax: 020 7067 1011
Email: info@housing.org.uk, www.housing.org.uk

National Register of Access Consultants

An independent register of accredited access auditors and access consultants who meet professional standards and criteria established by a peer review system.

70 South Lambeth Road, London SW8 1RL
Tel: 020 7840 0125, Textphone: 07921 700 089, Fax: 020 7840 5811
Email: info@nrac.org.uk, www.nrac.org.uk

Scottish Federation of Housing Associations

SFHA exists to support the work of housing associations and co-operatives in Scotland by providing services, advice and good practice guidance. Almost 200 housing associations in Scotland are members.

38 York Place, Edinburgh, EH1 3HU
Tel: 0131 556 5777, Fax: 0131 557 6028
Email: sfha@sfha.co.uk, www.sfha.co.uk

Welsh Federation of Housing Associations

The membership body for housing associations in Wales, with 67 members who manage/own 72 000 homes throughout Wales.

Norbury House, Norbury Road, Fairwater, Cardiff CF5 3AS
Tel: 029 2030 3160
Email: federation@welshhousing.org.uk, www.welshhousing.org.uk

Index

provision for 104
secondary doors 47
door pulls *see* pull handles
doors 7, 10–11, 12–13
balconies 23, 25
bedrooms 88, 89, 90, 102
circulation areas 46, 54–5, 57, 58–9, 61
clear opening widths 12, 54, 60, 92
communal areas 35
entrance halls 43, 44
gardens 46–9
hoists and 15, 80, 89, 93
lighting controls 36, 47, 102
opening angles 8, 12, 53, 55, 61, 89
protection 56, 92, 94
remote controls 34, 37, 92, 95
showers 80
storage facilities 56, 59, 72, 91
turning spaces 7, 42, 44, 45, 51–2, 59
see also bathroom doors; double-swing doors; effective clear widths; entrance doors; fire-resisting doors; garages; internal doors; inward-opening doors; opposing doors; outward-opening doors; secondary doors; self-closing doors; sliding doors
double-swing doors 55, 93–4
drawers 14, 73
dropped kerbs 18

see also charging facilities; controls; socket outlets
electricity main switches 103
emergency access, bathrooms and WCs 79, 92, 94
emergency communications 63, 102
emergency exits 46–9
emergency lighting 102, 106
entrance canopies 6, 27, 30, 31, 32
common detailing 20
refuse disposal facilities and 24
entrance doors 33–40
approach spaces 29, 33–4, 36–7, 39, 41–2, 45
common detailing 20
effective clear widths 33, 36, 39, 44
entrance landings and 31
mat clearance 43
security 33, 34, 35, 43, 45
turning spaces 7, 42, 44, 45
entrance halls 29, 41–5, 51–2, 53, 102
entrance landings 29, 31, 32
entrance ramps 30
entrances 6, 27–32
lighting controls 36, 102
signs 22, 36
telephone services 102, 104
turning spaces 7, 41–2, 44, 45
see also entry phones; external lighting
entry phone points 44, 45, 89, 90, 104
entry phones 42, 44, 45
accessibility 29, 33
door operators and 35
entrance landing areas 29, 32
locks and 34
positioning 38
selection 35
visual display 35, 39
external lighting 22, 29, 31, 36, 38
local authority standards 18
PIR (passive infrared) detectors 29, 31, 36, 38, 40, 49

floors
bathrooms 77, 83, 85, 86
cleaning 74, 83
garages 30
kitchens 74
see also mats
footpaths 20–2
frontages 19, 22
furniture layouts 14, 65, 96, 98